CLOCK SYNCHRONIZATION AND NAVIGATION IN THE VICINITY OF THE EARTH

CLOCK SYNCHRONIZATION AND NAVIGATION IN THE VICINITY OF THE EARTH

THOMAS B. BAHDER

Nova Science Publishers, Inc.
New York

For permission to use material from this book please contact us:
Telephone 631-231-7269; Fax 631-231-8175
Web Site: http://www.novapublishers.com

NOTICE TO THE READER

Library of Congress Cataloging-in-Publication Data
Available upon request.

ISBN 978-1-60692-114-2

Published by Nova Science Publishers, Inc. ✛ *New York*

Contents

Preface

Clock synchronization is the backbone of applications such as high-accuracy satellite navigation, geolocation, space-based interferometry, and cryptographic communication systems. The high accuracy of synchronization needed over satellite-to-ground and satellite-to-satellite distances requires the use of general relativistic concepts. The role of geometrical optics and antenna phase center approximations are discussed in high accuracy work. The clock synchronization problem is explored from a general relativistic point of view, with emphasis on the local measurement process and the use of the tetrad formalism as the correct model of relativistic measurements. The treatment makes use of J. L. Synge's world function of space-time as a basic coordinate independent geometric concept. A metric is used for space-time in the vicinity of the Earth, where coordinate time is proper time on the geoid. The problem of satellite clock syntonization is analyzed by numerically integrating the geodesic equations of motion for low-Earth orbit (LEO), geosynchronous orbit (GEO), and highly elliptical orbit (HEO) satellites. Proper time minus coordinate time is computed for satellites in these orbital regimes. The frequency shift as a function of time is computed for a signal observed on the Earth's geoid from a LEO, GEO, and HEO satellite. Finally, the problem of geolocation in curved space-time is briefly explored using the world function formalism.

Chapter 1

Introduction

In the past several decades, there has been a dramatic improvement in two technology areas: atomic clocks and lasers for free-space optical communications. Future spacecraft are envisioned as communicating by free-space laser links. The synergy of technological development in atomic clocks and free-space lasers will lead to unprecedented advancements in high-accuracy space-time navigation [1, 2], digital communications, geolocation [3–7], surveillance using space-based interferometers [8, 9], coherent distributed-aperture sensing at high frequencies [10–17], and cryptographic communication systems.

Accurate clock synchronization is the backbone of these systems. Consider the dependence of geolocation accuracy on clock synchronization. Consider geosynchronous satellites that receive a signal from an emitter of electromagnetic radiation located on the surface of the Earth. Assuming that the signals travel on the line-of-sight, the geometry leads to a maximum signal time difference of arrival between two satellites that is 19.64 ms, see Appendix A. The information on the difference of ranges, Δl, between the emitter and each satellite, is contained in the maximum time delay that is equal to 19.64 ms. If the clocks in two satellites are synchronized only to an accuracy of, say 10 ns, the order of magnitude in the position error of the emitter, Δx, is given by the fraction of the range difference:

$$\Delta x \sim \frac{10\text{ns}}{19.64\text{ms}} \Delta l \sim 3000\text{m} = 1.5\text{nm} \tag{1.1}$$

An improvement in the clock synchronization translates directly into an improvement in position accuracy. For example, an improvement in clock synchronization by three orders of magnitude will produce a position accuracy on the order of 3 m.

Applications such as a space-based interferometer, can have even more stringent requirements on clock synchronization. For example, a multi-satellite space-based interferometer (distributed aperture system) that operates at a wavelength λ will require accurate determination of relative satellite positions (nodes of the interferometer) to better than $\Delta x = \lambda$. For practical purposes, say we will need $\Delta x = \lambda/10$, which translates to a clock synchronization requirement $(\lambda/10c)$, where c is the speed of light. As an example of the stringent requirements on time synchronization, consider operating at 22 GHz, which is a frequency of interest to radio astronomers [18]. At this frequency, the position of the satellite nodes must be known accurately to $\lambda/(10c) = 1.4$ mm and time synchronization to 4.5 ps. See Table 1.1 for a range of values corresponding to different frequencies. For operating at optical wavelengths of 500 nm, the required position must be known to 50 nm and time must be synchronized to 0.16 fs. These numbers challenge and exceed the realm of possible time synchronization accuracy that is available today. However, the synergy between accurate clocks and optical free-space communication is expected to continue, so that hardware may soon support such stringent time synchronization requirements. New time synchronization schemes will then be required.

The emerging field of quantum information and quantum computation [19, 20] has potential to produce new ultra-precise clock synchronization protocols. In fact, several clock synchronization schemes have been proposed based on quantum mechanical concepts [21–30]. However, most of these schemes have neglected real features of the clock synchronization problem that are essential for real-life applications: *the clocks to be synchronized are in relative motion, and at varying gravitational potentials.* The fact that clocks are affected by their motion and by the gravitational potential are basic concepts that have their origin in Einstein's special and general relativity theory. The examples of the required accuracy of clock synchronization cited above, together with the magnitudes of relativistic effects on satellites (see Section IV, subsection E), show that the required accuracies can only be met by theories that take into account the effect of gravitational potential differences and relative motion. If synchronization of clocks is to be achieved using quantum information concepts, then certain features, such as relative motion of clocks and effects of gravitational potential, must be incorporated in the quantum information approach to clock

Table 1.1. Frequency f, wavelength λ, and $\lambda/(10c)$ time synchronization is shown for various operating regimes of an interferometer.

f	λ	$\lambda/(10c)$ time synchronization
1500 kHz	200 m	66 ns
20 MHz	15 m	5 ns
120 MHz	2.5 m	0.83 ns
20 GHz	1.5 cm	5.0 ps
60 GHz	5 mm	1.6 ps
30 THz	10 μ m	3.3 fs
500 THz	500 nm	0.16 fs

synchronization.

This chapter addresses the theoretical problem of clock synchronizing and syntonization (making two clocks run at the same rate), or correlating time on clocks on satellite platforms that are orbiting Earth, or that are near-Earth. However, Lorentz transformations between two systems of coordinates in relative motion show that space and time are really interwoven. The problem of clock synchronization is really part of the more general problem of navigation in space-time [1, 2]. For example, in the Global Positioning System (GPS), a user's receiver determines three spatial coordinates as well as time [2]. Therefore, in this article we will focus on the complete problem of navigation in space-time from the point of view of curved space-time, such as is invoked in general relativity. The features that relativity deals with, motion of clocks and effect of gravitational potential, are also features that must also be included in any classical or quantum theory of space-time navigation. We use the fact that space-time is described by a four dimensional metric, but for the most part we do not explicitly use the field equations of general relativity. In this sense our discussion is not restricted to general relativity.

In this article, several considerations are stressed in the space-time navigation problem. First, space-time navigation is based on real measurements made by real physical devices. Real measurements are (spatially and temporally) local quantities that are invariants under change of space-time coordinates (see Section VI). Historically, due to a lack of accurate measurements over large distances (such as spacecraft to ground) measurement theory has not played a large role in relativity theory. On the other hand, measurements are at the core of quantum theory. Perhaps this point of intersection between relativity and quantum theory will help clarify how to augment quantum information theory with relativistic ideas.

Chapter 2

Hardware Time, Proper Time, and Coordinate Time

A clock is a physical device consisting of an oscillator running at some angular frequency ω, and a counter that counts the cycles. The period of the oscillator, $T = 2\pi/\omega$, is calibrated to some standard oscillator. The counter simply counts the cycles of the oscillator. Since some epoch, or the event at which the count started, we say that a quantity of time equal to NT has elapsed, if N cycles have been counted.

In this article I distinguish between three types of time: hardware time τ^*, proper time τ, and coordinate time t. Hardware time is associated with a real physical device that keeps time, which I call a *hardware clock*. Specifically, *hardware time* $\tau^* = NT$ is the time kept by a hardware clock, and is given in terms of the number of cycles N counted by the device. Two hardware clocks will differ in the elapsed time that they indicate between two events, because no two devices are exactly the same. Furthermore, heating, cooling, and vibration typically affects real devices, and consequently the value of the elapsed hardware time registered on a real hardware clock can vary for these reasons.

Proper time is an idealized time interval occurring in the theory of relativity. We imagine that there exists an ideal clock (oscillator plus counter) that is unaffected by temperature or vibration. However, based on Einstein's theory of relativity [31,32], the ideal clock is affected by gravitational fields, acceleration and velocities. According to Einstein's general theory of relativity, gravitational fields, acceleration and velocities affect all physical processes, and hence these effects are associated with the geometry of space and time. A basic tenet of the theory is that between any two events that are infinitesimally separated in space-time by dx^i, $i = 0, 1, 2, 3$, there exists an invariant quantity ds called the space-time interval [33]

$$ds^2 = -g_{ij}dx^i dx^j \qquad (2.1)$$

where g_{ij} is the metric of the 4-dimensional space-time. In general relativity, if the two events are time-like, then, $ds^2 > 0$, and the events are "separated farther in time than in space". We interpret $ds = cd\tau$, where $d\tau$ is the proper time elapsed on an ideal clock that moves between these two events.

The definition of an ideal clock is one that keeps proper time intervals. More specifically, if a clock moves from point P_1 to point P_2 on a 4-dimensional world line given by coordinates $x^i(u)$, $i = 0, 1, 2, 3$, for $u_1 \leq u \leq u_2$, where u is some parameter such that $P_1 = \{x^i(u_1)\}$ and $P_2 = \{x^i(u_2)\}$, then the proper time interval $\Delta\tau$ between these two events is given by the path integral in the space-time:

$$\Delta\tau = \int_{u_1}^{u_2} \sqrt{-g_{ij}\frac{dx^i}{du}\frac{dx^j}{du}}\, du = \int_{x_1^0}^{x_2^0} \sqrt{-g_{ij}\frac{dx^i}{dx^0}\frac{dx^j}{dx^0}}\, dx^0 \qquad (2.2)$$

where the second integral has been parametrized by the coordinate time length $x^0 = ct$, where t has units of time. From Eq. (2.2), it appears that the proper time interval $\Delta\tau$ depends on the metric of the space-time, g_{ij}, and on the 4-velocity of the clock, dx^i/ds. But in fact, $\Delta\tau$ is a geometric quantity that depends on the path in space-time, and is independent of the coordinates used to compute $\Delta\tau$. For the purpose of applications, Eq. (2.2) provides a functional relation between the proper time interval measured on an ideal clock, $\Delta\tau$, and the path which the clock traversed in 4-dimensional space-time. Under certain conditions, Eq. (2.2) can provide a relation between elapsed proper time $\Delta\tau$ and coordinate time interval, $\Delta x^0 = x_2^0 - x_1^0$,

$$\Delta\tau = f_1(x_1^0, x_2^0) \qquad (2.3)$$

which depends on the coordinate time x_1^0 and x_2^0 of events P_1 and P_2, respectively. However, in general $\Delta\tau$ depends on the whole path $x^i(u)$, and not just on the end points of the path.

A good, real clock, which we will call a hardware clock, is believed to provide an approximate measure of elapsed proper time. Therefore, for a given real hardware clock, there is some functional relation between the elapsed proper time, $\Delta\tau$, and hardware time, $\Delta\tau^*$:

$$\Delta\tau^* = f_2(\Delta\tau) \qquad (2.4)$$

where the function $f_2()$ has some stochastic aspects and is hardware dependent.

A peculiarity of the theory of relativity is that the coordinate time, x^0, as well as the other coordinates x^α, $\alpha = 1, 2, 3$, are not directly measurable quantities [34]. The coordinates are only mathematical constructs, and cannot be measured directly. Events occurring in space-time have coordinates associated with them. Two events are labelled with coordinates $P_1 = \{x_1^i\}$ and $P_2 = \{x_2^i\}$ even though these quantities are not directly measurable. However, the theory allows us to relate the difference of coordinate time, $x_2^0 - x_1^0$, between two events, to the proper time interval, by the path integral in Eq. (2.2). Of course, as mentioned previously, neither of these quantities are measurable directly, instead, only hardware time intervals, $\Delta \tau^*$, are measured directly from hardware clocks. Everything else is calculated.

The relations in Eq. (2.2)–(2.4) permit, under some circumstances, the relation of measured hardware times $\Delta \tau^*$ to coordinates times Δx^0, which enter into the theory or relativity.

2.1. Synchronization versus Syntonization

The relation between proper time $\Delta \tau$ and coordinate time is such that they may "run at different rates". This is clearly the case when the metric of the space-time is not a constant over the integration path. For example, a good clock may measure proper time intervals accurately, however, because of Eq. (2.2), this clock will run at a rate that differs from coordinate time x^0, because $d\tau/dx^0 \neq$ constant. While coordinate time is a global coordinate quantity, valid (almost) everywhere in the space-time, the proper time interval depends on the world line of the clock. For example, for an ideal clock that is stationary (has constant spatial coordinates x^α, $\alpha = 1, 2, 3$) with world line $x^i = (x^0, x^1, x^2, x^3)$, for $x_o^0 < x^0 < x_o^0 + dx^0$, the proper time interval is $d\tau = \sqrt{-g_{00}}dx^0$. Therefore, the rate of proper time, $d\tau/dx^0$, depends on position thorough g_{00}.

Consider now two ideal clocks at the same location and assume that these two clocks are synchronized to read the same starting time at some epoch, or starting event. Next move the clocks apart (hence they travel on different world lines) and then bring them together once again to a common location. On comparing the times on these clocks, we find that different amounts of proper time have elapsed on each of them. In other words, proper time ran at different rates for each of the clocks. We say that the two clocks were not syntonized (i.e., they did not run at the same rate) since, when they were brought back together a different amount of proper time elapsed on each clock. The rate of each clock can be compared at any instant to the underlying coordinate time (which is a globally defined quantity), by using Eq. (2.2), and in this way the time on one clock can be compared to the time on the other clock. In Section IX, we will find that the most serious problem for practical satellite applications is the syntonization of clocks.

Chapter 3

Choice of a Physical Theory

A theory must be chosen as the basis for navigation and clock synchronization. The theory must deal with two areas: the space-time in which all events occur, and the propagation of electromagnetic fields in this space-time. While these two areas are coupled in the sense that electromagnetic fields are sources for the curvature of space-time (within general relativity), we will take these two areas separately. More specifically, we will neglect the (very small) effect of the electromagnetic field creating space-time curvature. We will assume that all curvature is caused by the presence of mass. For applications in the vicinity of the Earth, this is an excellent approximation.

The most common theory to choose is general relativity, as developed by Einstein [31,32]. As has already been discussed, this theory includes the effects of motion on clocks and also the effects of gravitational potential on clocks. In fact, there are several variants of relativistic theories of gravitation. However, Einstein's general theory of relativity has so far passed all physical tests [35,36], and we shall subscribe to it in this report. Einstein's general theory of relativity can be divided into two parts: first, the theory assumes that there exists a space-time metric g_{ij}, given by Eq. (2.1), which relates proper time $d\tau = ds/c$ to the space-time geometric interval ds, and coordinate difference dx^i between two events. This idea is quite general, and is really an embodiment of the principle of equivalence [35,36]. The principle of equivalence essentially states that gravitational mass (in Newton's universal law of gravitation) and inertial mass (in Newton's second law of motion) are the same quantity. Roughly speaking, the equivalence principle says that two small test bodies of different mass will fall along the same geodesic path, i.e., the same distance in the same time. This principle has been extensively tested [35, 36], and is the basis of essentially all metric-based theories of gravity, because they are based on geometric ideas embodied in Eq. (2.1). Physicists today generally agree that a theory of gravity should be a metric theory in a pseudo-Riemannian curved space-time with metric of the form given by Eq. (2.1).

The second part of Einstein's general relativity theory consists of the field equations [31, 32]

$$G^{ij} = -\kappa T^{ij} \tag{3.1}$$

where $\kappa = 8\pi G/c^2$, where G is Newton's gravitational constant, and c is the speed of light in vacuum. The field Eqs. (3.1) relate the matter distribution, given by the stress energy tensor T^{ij}, to the effect that this matter has on space-time via the Einstein tensor,

$$G_{ij} = R_{ij} - \frac{1}{2} g_{ij} R \tag{3.2}$$

where the Ricci tensor, $R_{ij} = R^k_{ijk}$, is related to the Riemann tensor

$$R^i_{jkm} = \Gamma^i_{jm,k} - \Gamma^i_{jk,m} + \Gamma^a_{jm} \Gamma^i_{ak} - \Gamma^a_{jk} \Gamma^i_{am} \tag{3.3}$$

and the affine connection

$$\Gamma^i_{jk} = \frac{1}{2} g^{il} \left(g_{jl,k} + g_{kl,j} - g_{jk,l} \right) \tag{3.4}$$

is related to the metric g_{ij}. So the field Eqs. (3.1) are a set of equations for the components of the metric tensor field g_{ij}. In Eq. (3.4), ordinary partial derivatives with respect to the coordinates are indicated by commas.

The field equations of Einstein, given in Eq. (3.1), have only been solved and tested in a limited number of cases. Consequently, the field equations are on a less-firm footing than the equivalence principle. Fortunately, most of the applications of navigation and clock synchronization in a curved space-time rely only on the fact that space-time is a metric theory, given by Eq. (2.1). Therefore, our conclusions below regarding navigation and clock synchronization transcend general relativity. Specifically, our conclusions are based on the assumed-correctness of the equivalence principle.

3.1. Electromagnetic Waves

The electromagnetic field plays a central role in experiments and applications. In technology applications, all information is currently carried by travelling electromagnetic fields. Inter-satellite links, and ground to satellite links are all done using electromagnetic radiation fields. Time transfer between stations on the ground and satellites is done by electromagnetic fields. Since electromagnetic fields paly such a key role, in this subsection we briefly outline the equations governing the propagation of electromagnetic waves, namely the Maxwell equations in flat space-time, and their generalization to curved space-time.

Electromagnetic fields in a medium, such as air, a dielectric, or a magnet, are described by the macroscopic Maxwell's equations, and in SI units, in *flat space-time*, these equations take the form

$$\text{div}\,\mathbf{D} = \rho \tag{3.5}$$

$$\text{div}\,\mathbf{B} = 0 \tag{3.6}$$

$$\text{curl}\,\mathbf{H} = \mathbf{J} + \frac{\partial \mathbf{D}}{\partial t} \tag{3.7}$$

$$\text{curl}\,\mathbf{E} = -\frac{\partial \mathbf{B}}{\partial t} \tag{3.8}$$

where $\mathbf{E}$ and $\mathbf{B}$ are the electric field and magnetic induction (or magnetic field), respectively, and $\mathbf{D}$ and $\mathbf{H}$ are the electric displacement field and magnetic intensity, respectively. In a medium, the fields $\mathbf{E}$ and $\mathbf{D}$, and the fields $\mathbf{B}$ and $\mathbf{H}$ are related by constitutive relations. In a vacuum, these fields are simply related by:

$$\mathbf{D} = \varepsilon_o\,\mathbf{E} \tag{3.9}$$

$$\mathbf{B} = \mu_o\,\mathbf{H} \tag{3.10}$$

where ε_o and μ_o are the permittivity and permeability of vacuum, respectively. In an isotropic (but not necessarily homogeneous) medium such as the Earth's atmosphere, the constitutive equations are

$$\mathbf{D} = \varepsilon\,\mathbf{E} \tag{3.11}$$

$$\mathbf{B} = \mu\,\mathbf{H} \tag{3.12}$$

where ε and μ are the permittivity and permeability of the medium, respectively.

In a curved space-time, the physics of electromagnetic wave propagation has been less explored [32, 37–41]. However, it is known that the gravitational field scatters and diffracts electromagnetic waves, and that the plane of polarization of an electromagnetic wave is rotated as the wave propagates through a gravitational field. In general, a gravitational field affects electromagnetic wave propagation similarly to a dispersive medium [32]. For weak fields, such as exist in the vicinity of the earth, these gravitational effects are smaller than the dispersive effects due to the atmosphere (at low altitude). The general equations that govern electromagnetic wave phenomena in curved space-time, in the presence of a dielectric are given by [40]

$$F_{ij,k} + F_{jk,i} + F_{ki,j} = 0 \qquad (3.13)$$

and

$$H^{ik}{}_{;k} = J^i \qquad (3.14)$$

where F_{ij} and H^{ik} are two antisymmetric tensor fields, the comma indicates partial differentiation with respect to the coordinates and the semicolon indicates covariant differentiation with respect to the coordinates. All Roman indices take values $i = 0, 1, 2, 3$. The two fields are related by a constitutive relation

$$\sqrt{-g}\sqrt{-\gamma}H^{ik} = c^2 \left(\frac{\varepsilon}{\mu}\right)^{\frac{1}{2}} \gamma^{ia}\gamma^{kb}F_{ab} \qquad (3.15)$$

The constitutive relation in Eq (3.15) assumes that the medium is isotropic, but not necessarily homogeneous, so the permittivity ε and permeability μ are both functions of position. In Eq (3.15), we have used the definitions of the contravariant components of the metric tensor g^{ij} and the effective metric

$$\gamma^{ij} = g^{ij} - (n^2 - 1)u^i u^j \qquad (3.16)$$

where n is the scalar index of refraction given by

$$n^2 = \frac{\varepsilon\mu}{\varepsilon_o\mu_o} \qquad (3.17)$$

where $g = \det g_{ij}$ and $\gamma = \det \gamma_{ij}$, u^i is the local 4-velocity of the medium in our system of coordinates, and J^i is the 4-current density. In the proper frame of reference of the medium (where the medium is at rest), the field tensors take the simple forms:

$$F_{ik} = \begin{pmatrix} 0 & -E_x & -E_y & -E_z \\ E_x & 0 & cB_z & -cB_y \\ E_y & -cB_z & 0 & cB_x \\ E_z & cB_y & -cB_x & 0 \end{pmatrix} \tag{3.18}$$

and

$$H_{ik} = \begin{pmatrix} 0 & -cD_x & -cD_y & -cD_z \\ cD_x & 0 & H_z & -H_y \\ cD_y & -H_z & 0 & H_x \\ cD_z & H_y & -H_x & 0 \end{pmatrix}, \tag{3.19}$$

the current density is $J^i = (\rho_o c, J^\alpha)$, where ρ_o is the proper charge density and J^α is the current. In this proper frame of reference of the medium, Eqs. (3.13)–(3.14) reduce to Eqs (3.5)–(3.8), and the constitutive Eq. (3.15) reduces to relations in Eq. (3.11)–(3.12).

If the medium is vacuum, then $n = 1$ and $\gamma^{ij} = g^{ij}$, and the tensors F_{ik} and H_{ik} are not independent, instead they differ only by trivial constants of the vacuum.

Equations (3.13)–(3.14) govern the propagation of electromagnetic fields in the presence of a medium in a gravitational field. In general, in the vicinity of the Earth, the dispersive effects of the medium are larger than those of the gravitational field. Due to the complexity of these equations, they have not been explored in detail. Only several treatments have been attempted, see for example Refs. [38, 39, 41]. The Eqs. (3.13)–(3.14) or (3.5)–(3.8), form the basis for applications such as time transfer, clock synchronization and communication. These equations are valid in a system of coordinates (frame of reference) that is in arbitrary motion. In particular, these equations describe propagation of electromagnetic waves, and the reception and transmission properties of antennas in the radio portion of the spectrum.

3.2. The Geometrical Optics Approximation

When discussing precise measurements, it is important to state precisely the theory and approximations used. Up to this point in time, the geometric optics approximation has been universally used in discussions of time transfer, usually without stating its use, and without examining the limitation of this approximation. Below, we state the geometric optics approximation and how it enters into time transfer ideas.

The geometric optics approximation consists of the assumption that the wavelength of the travelling electromagnetic wave, λ, is much smaller than the linear dimension l of all objects (length scales) in the physical problem under consideration [32]:

$$\lambda << l \qquad (3.20)$$

In physical terms, the limit of short wavelength waves (geometrical optics limit) corresponds to waves that travel along straight lines, so that diffraction (e.g., bending of waves around boundary edges) is absent. As an example of geometric optics at work, consider the visible shadows cast on the ground by objects in the path of the light from the sun to ground. The light travels at approximately straight lines. However, if one looks very closely near the edge of the shadow, the boundary between light and dark areas is not sharp, and this is where geometrical optics shows its limitation–there is diffraction, or bending of the rays around sharp edges of an opaque material.

In the literature, statements are often made in the context of special relativity theory that in flat space-time "light travels along a straight line", or in curved space-time, that "light travels along a geodesic". Both of these statements are true only within the context of the geometrical optics limit [42–45]. In recent years, there has been limited work to explore the nature of travelling electromagnetic waves in a curved space-time, going beyond the geometrical optics approximation [38, 39, 41–45]. The gravitational field creates complex effects such as diffraction of the electromagnetic wave and rotating its polarization.

3.3. Signal Detection and Use of Antenna Phase Center

In precision measurements, it is important to have a clear concept of the point from which radiation emanates and the point at which the radiation is detected. Usually, such a discussion makes implicit use of the geometric optics approximation.

One place where the geometric optics approximation is not well-satisfied is for real (radio frequency) antennas, because antennas are efficient at receiving and transmitting radiation at wavelengths that are comparable to the antenna size, so Eq. (3.20) is not well-satisfied. The desire to continue to use the (very convenient) geometric optics approximation forces us to invoke the concept of antenna phase center in precise time transfer or navigation applications. We then imagine that there is a unique emission point, from which radiation emanates, and a unique reception point, where the radiation is absorbed.

The antenna phase center is defined as the apparent point from which radiation emanates (or is absorbed). In the far-field radiation region, for one vector component of the electric field of antenna radiation, and for one polarization and one frequency ω, the electric field can be expressed as

$$\mathbf{E} = \mathbf{u} E(\theta, \phi) e^{\psi(\theta, \phi)} \frac{e^{ikr}}{r} e^{-i\omega t} \qquad (3.21)$$

where the vector $\mathbf{u}$ is a real polarization unit vector, $E(\theta, \phi)$ is the (real) electric field amplitude, $\psi(\theta, \phi)$ is the (real) phase, r is the distance from the antenna, and $i = \sqrt{-1}$. If a point can be found such that $\psi(\theta, \phi)$ is independent of θ and ϕ, i.e., independent of the direction of propagation, then this point is the antenna phase center. For most practical antennas, no such point exists [46,47]. The reception of an antenna is related to its transmission properties, with the same phase center, by reciprocity relations [47,48].

In the receive mode, the open-circuit voltage V induced in a receiving antenna [47–50] is given by the scalar product between the radiation field from a given satellite, $\mathbf{E}$, and the vector effective length $\mathbf{h}(\mathbf{n})$:

$$V = \text{Re}\{ \mathbf{h}(\mathbf{n}) \cdot \mathbf{E} \} \qquad (3.22)$$

where Re takes the real part of a complex expression, $\mathbf{E}$ is the electric radiation field at the receiving antenna, and $\mathbf{h}(\mathbf{n})$ is the receiving antenna vector effective length (sometimes called the effective height), which is a complex vector quantity that characterizes the electromagnetic wave phase relationship of the antenna in receive and transmit mode. The receive and transmit modes are related by reciprocity relations [47–50]

Despite the limitations of the concept of antenna "phase center", this concept is routinely invoked in practice in real antenna systems. The practical complication is that the antenna "phase center" is not a fixed point, but its position depends on electromagnetic wave frequency, ω, polarization vector $\mathbf{u}$, and direction of travel with respect to the antenna (both in receive and transmit modes). In other words the phase center position is not fixed, instead it varies with ω and $\mathbf{u}$ over some range of coordinate values Δx that is on the order of a wavelength of the radiation, $\Delta x \sim \lambda$. In other words, the point at which an antenna receives a signal is only precisely defined (in geometrical optics) to within a distance $\Delta x \sim \lambda$. Consequently, when one antenna transmits radiation and another antenna receives radiation, the effective distance between these two antennas can vary with frequency and polarization of the radiation, and with the relative orientation of the antennas. When the relative orientation of one satellite changes with respect to another satellite due to their relative motion, their effective separation changes due to change of relative orientation of their antennas, in addition to a real change of distance between them. Clearly, the concept of a single emission point and a single reception point is fuzzy on the order of the scale of a wavelength at both transmit and receive end points.

In conclusion, the apparent distance between antennas changes with frequency, polarization, and orientation. In a case where precise navigation is to be carried out using radio frequencies, these effects must be taken into account, so that an accuracy of better than one wavelength of the radiation can be achieved.

As an alternative to using radio frequencies, we can transition to satellites using optical frequencies, which have considerably shorter wavelengths. The transition to optical free-space communications in satellites, with smaller wavelengths on the order of $10 - 60$ nm, has the advantage that we can invoke the geometric optics approximation, and suffer $\Delta x \sim \lambda$ errors that are much smaller, because of the much smaller optical wavelengths. For example, for optical wavelengths in the range 10 nm to 60 nm, position errors are comparable to the wavelength, and this should be compared with radio frequencies, of say, 1 MHz to 10 GHz, with wavelength range of 3 cm to 300 m.

Actually, at optical frequencies, Eq. (3.22) is not the physical mechanism that is responsible for detecting electromagnetic fields. Instead, detectors of electromagnetic radiation work either as bolometers or quantum detectors. Bolometric detectors are based on the pyroelectric effect, which produces a change of dielectric polarization with increase of temperature, due to absorption of electromagnetic radiation [51]. The polarization change is detected electrically.

The second common mode of detecting optical radiation is based on a quantum mechanical effect. For example, in a semiconductor material a photon (quantum) of the electromagnetic field is absorbed by an electron, and the electron makes a transition from a valence band quantum state to a conduction band state. The electron in the conduction band is then detected electrically. In either case, the minimum volume that is needed for detection is roughly of the dimensions of a wavelength of the optical radiation, so similar criteria apply (to that of radio frequency) for accuracy of the point of absorption.

As discussed in the Introduction, we believe that future satellite systems will have optical links. Such satellites may have an interferometer that operates in the radio frequency portion of the spectrum, but the navigation (positioning) and time synchronization will be done optically. Consequently, the geometric approximation will be useful because of the (relatively) small optical wavelengths compared to radio frequency wavelengths. Therefore, we will freely use theory that assumes that electromagnetic waves travel on geodesics in curved space-time, which is the geometric optics approximation.

Chapter 4

World Function of Space-Time

The geometric optics approximation is valid for the applications addressed in this report. Within the geometric optics approximation we can say that electromagnetic waves travel on geodesics in space-time. This is a purely geometric statement: the emission and reception events of a signal are connected by a null geodesic. Geodesics in the space are determined by the metric of the space-time.

A useful quantity to deal with measurements in space-time is the world function Ω. The world function is simply one-half the square of the space-time interval (see Eq. (2.1)), measured along the geodesic connecting two points [1, 2, 31, 32].

The world function was initially introduced into tensor calculus by Ruse [52, 53], Synge [54], Yano and Muto [55], and Schouten [56]. It was further developed and extensively used by Synge in applications to problems dealing with measurement theory in general relativity [31]. The world function has generally received little attention in the literature, so we provide a detailed definition here. Consider two points, P_1 and P_2, in a general space-time, connected by a unique geodesic path Γ given by $x^i(u)$, where $u_1 \leq u \leq u_2$. A geodesic is defined by a class of special parameters u' that are related to one another by linear transformations $u' = au + b$, where a and b are constants. Here, u is a particular parameter from the class of special parameters that define the geodesic Γ, and $x^i(u)$ satisfy the geodesic equations

$$\frac{d^2 x^i}{du^2} + \Gamma^i_{jk} \frac{dx^j}{du} \frac{dx^k}{du} = 0 \qquad (4.1)$$

The world function between P_1 and P_2 is defined as the integral along Γ:

$$\Omega(P_1, P_2) = \frac{1}{2}(u_2 - u_1) \int_{u_1}^{u_2} g_{ij} \frac{dx^i}{du} \frac{dx^j}{du} \, du \tag{4.2}$$

The value of the world function has a geometric meaning: it is one-half the square of the space-time distance between points P_1 and P_2. Its value depends only on the eight coordinates of the points P_1 and P_2. The value of the world function in Eq. (4.2) is independent of the particular special parameter u in the sense that under a transformation from one special parameter u to another, u', given by $u = au' + b$, with $x^i(u) = x^i(u(u'))$, the world function definition in Eq. (4.2) has the same form (with u replaced by u').

The world function is a two-point invariant in the sense that it is invariant under independent transformation of coordinates at P_1 and at P_2. Consequently, the world function characterizes the geometry of the space-time. For a given space-time, the world function between points P_1 and P_2 has the same value independent of the coordinates that are used. As a simple example of the world function for Minkowski space-time, consider

$$\Omega(x_1^i, x_2^j) = \frac{1}{2} \eta_{ij} \Delta x^i \Delta x^j \tag{4.3}$$

where η_{ij} is the Minkowski metric with only non-zero diagonal components $(-1, +1, +1, +1)$, and $\Delta x^i = (x_2^i - x_1^i)$, $i = 0, 1, 2, 3$, where x_1^i and x_2^i are the coordinates of points P_1 and P_2, respectively. Up to a sign, the world function gives one-half the square of the geometric measure (the interval) in space-time. Calculations of the world function for specific space-times can be found in Refs. [1, 2, 31, 57–59] and application to Fermi coordinates in Synge [31] and Gambi et al. [60].

The world function for the Schwarzschild metric, linearized in small parameter $GM/c^2 r$, is given by [1]:

$$\Omega(x_1^i, x_2^j) = \frac{1}{2} \eta_{ij} \Delta x^i \Delta x^j + \frac{GM}{c^2} \left[|\mathbf{x}_2 - \mathbf{x}_1| + \frac{c^2 \Delta t^2}{|\mathbf{x}_2 - \mathbf{x}_1|} \right] \log \left(\frac{\tan(\frac{\theta_1}{2})}{\tan(\frac{\theta_2}{2})} \right)$$
$$+ \frac{GM}{c^2} |\mathbf{x}_2 - \mathbf{x}_1| (\cos\theta_1 - \cos\theta_2) \tag{4.4}$$

where $c\Delta t \equiv x_2^0 - x_1^0$, and θ_1 and θ_2 are defined by

$$\cos\theta_a = \frac{\mathbf{x}_a \cdot (\mathbf{x}_2 - \mathbf{x}_1)}{|\mathbf{x}_a| |\mathbf{x}_2 - \mathbf{x}_1|}, \quad a = 1, 2 \tag{4.5}$$

The world function in Eq. (4.4) assumes a spherical Earth ($J_2 = 0$), see Eq. (8.2). For most applications dealing with delays of electromagnetic wave propagation, this approximation is sufficient. A more accurate calculation of the world function for Schwarzschild metric is given by Buchdahl and Warner [59]. Including the small effect of the oblateness of the Earth will require additional computations and is left for future work. For a pedagogic discussion of applying the world function to space-time navigation, see Ref. [1].

4.1. Navigation in Curved Space-time

A satellite must be localized with respect to some system of coordinates. The satellite can receive signals from four electromagnetic beacons and use the measurements to solve for its position. Alternatively, the satellite can send out electromagnetic signals that are received at four locations (e.g., on the Earth surface) , and these four quantities can be used to compute the satellite position at emission time.

In either case, the equations for navigation in a curved space-time can be formulated in a covariant, and invariant way (independent of coordinates) using the world function [1]. For example, consider a satellite with unknown coordinates $x_o^i = (t_o, \mathbf{x}_o)$, $i = 0, 1, 2, 3$, which we want to locate with respect to four electromagnetic beacons having coordinates $x_a^i = (t_a, \mathbf{x}_a)$, $a = 1, 2, 3, 4$. Assume that the satellite simultaneously receives the four signals at the space-time event with coordinates x_o^i. We must solve the four equations given by

$$\Omega(x_o^i, x_a^j) = 0, \quad a = 1, 2, 3, 4 \tag{4.6}$$

for the unknown satellite coordinates, $x_o^i = (t_o, \mathbf{x}_o)$, in terms of the known emission event coordinates, $x_a^i = (t_a, \mathbf{x}_a)$. These equations state that the emission and reception events are connected by null geodesics. In addition to Eq. (4.6), the appropriate causality conditions $t_o > t_a$, for $a = 1, 2, 3, 4$, must be added. The set of relations in Eq. (4.6) are manifestly covariant and invariant due to the transformation properties of the world function under independent space-time coordinate transformations at point P_1 and at P_2. Equations (4.6) neglect atmospheric effects, although these could be included in future work.

From the definition of the world function, the intrinsic limitations of navigation in a curved space-time are evident: the world function $\Omega(P_1, P_2)$ must be a single-valued function of P_1 and P_2. In general, if two or more geodesics connect the points P_1 and P_2, then $\Omega(P_1, P_2)$ will not be single-valued and the set of equations in Eq. (4.6) may have multiple solutions or no solutions. Such conjugate points P_1 and P_2 are known to occur in applications to planetary orbits and in optics [31]. However, when the points P_1 and P_2 are close together in space and in time and the curvature of space-time is small, we expect the world function to be single valued and the solution of Eq. (4.6) to be unique. Therefore, navigation in curved space-time is limited by the possibility of determining a set of four unique null geodesics connecting four emission events to one reception event. In the vicinity of the Earth, there is no ambiguity, due to the weak gravitational field. However, in the case of strong gravitational fields, as may exist in the vicinity of a black hole, or when the (satellite) radio beacons are at large distances from the observer in a space-time of small curvature, navigation by radio beacons may not be possible in principle. In such cases, one may have to supplement radio navigation by inertial techniques; see for example the discussion by Sedov [61].

Chapter 5

Physical Measurements

In order to extract information from a quantum mechanical system, a measurement has to be performed. In a quantum mechanical theory, the measurement process plays a central role. The quantum mechanical measurement process is imagined as an interaction that occurs between a classical, macroscopic apparatus and the quantum system [19, 20, 62]. Much effort has been expended on detailed investigations of the measurement process.

In contrast, in the general theory of relativity, comparatively little attention has been devoted to the role of the measuring process. In part, this is due to the fact that really high-accuracy measurements were not carried out due to the associated technological difficulties. However, the presence of high accuracy clocks in space make high accuracy measurements a reality.

5.1.   Observations and Measurements

In a real laboratory experiment, the process of taking data consists of recording information that is observed on physical meters, dials and gauges. For example, the reading on a volt meter can be observed at some point in time and space. The reading on the volt meter is an example of an *observation*. The observation occurs at some point in space and time (an event), marked by coordinates in some 4-dimensional system of space-time coordinates. The space-time coordinates of an observation may not be known to the person that is making the observation.

Physically useful observations usually consist of coincidences of two (or more) things occurring at the same place at the same time. As an example of

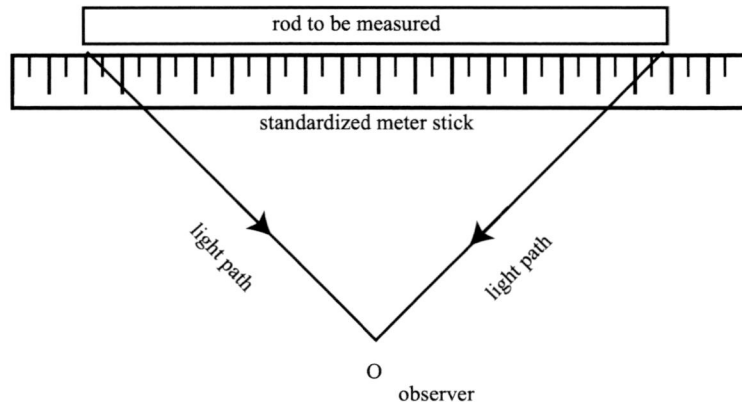

Figure 5.1. The standardized meter stick is shown next to the rod to be measured.

an observation, consider walking on a side walk, and noting the time on your watch when passing a crack in the sidewalk. The time noted is an observation of a coincidence: the position of my watch with the position of the crack in the sidewalk. This coincidence occurred at a space-time event, which has some coordinates. The observation is idealized as occurring at a given point in space-time, and each situation must be analyzed whether the "region of observation" satisfies, for the given application, the accuracy requirement of the observation as occurring at a point.

In contrast to an observation, a *measurement* is an observation in which a comparison is made. For example, consider an a.c. electric current that is fed into a circuit where the phase of the incoming current is compared (*measured*) to the phase of a reference a.c. current. The measurement is the phase difference between these two currents. The measurement occurs at a specific space-time event, with definite coordinates. In the real world, the event is not a point, but often, to a sufficient degree of accuracy it can be modelled as occurring at one space-time point. Whether any given measurement can be regarded as occurring at one point in space-time depends on the required accuracy, and must be analyzed on a case by case basis.

As another example of a measurement process, consider measuring a rod, by use of a standardized meter stick. Light from the ends of the rod comes to our eyes, along with light from the graduated scale on the standardized meter stick. The (simultaneous) event of light entering our eye from the left and right sides of the rod and meter stick constitute an observation, and because it is a comparison, it is also a measurement. See Fig. 5.1. Measurements are a subset of observations.

Measurements are dimensionless ratios: the thing measured is compared to a standard. Furthermore, measurements are invariant quantities. In general relativity theory, the tetrad formalism treats measurements as invariant quantities.

5.2. Tetrad Formalism

Consider two observers that are making spectral measurements on light from the same star. Assume that the two observers are in relative motion, but that at the instant of measurement, they are located at the same point in space. At this point, the observer that is moving toward the star may measure predominantly blue light emitted from the star. On the other hand, the other observer that is travelling away from the star may measure predominantly red light. So two measurements at the same place at the same time lead to different results. (There are other types of measurements that may produce identical results for the two observers.) In the previous section, we stated that measurements are invariant quantities. In what sense then are measurements invariant?

In connection with measurements, there are two types of transformations that must be considered. First, the global coordinates in the space, x^i, can be transformed to new coordinates, say using a transformation such as

$$x^i \rightarrow y^i = f^i(x^k) \tag{5.1}$$

where $f^i()$ are a set of transformation functions. Since measurements are scalar quantities (see below), they are always invariant under the generalized coordinate transformations of the type in Eq. (5.1).

The second type of transformation that must be considered in connection with measurements is that two observers have different world lines and consequently different tetrad basis vectors onto which they project electromagnetic fields. The projection onto the tetrad is the measurement process. Real measurements are local quantities, and they can be compared when two observers are colocated at the same space-time point, see Figure 5.2. Transformations from the tetrad basis $\lambda^i_{(a)}$, $a = 0, 1, 2, 3$, of one observer to the tetrad basis of

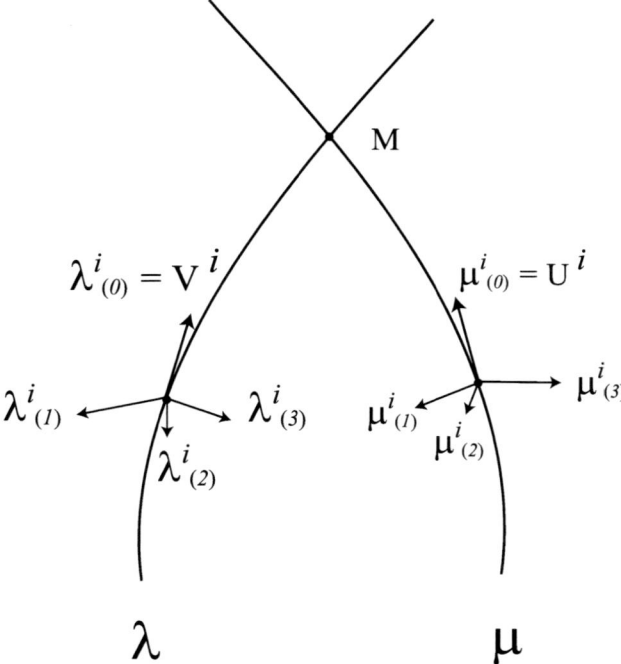

Figure 5.2. The world lines of two observers, λ and μ, are shown with their respective tetrads, $\lambda^i_{(\alpha)}$ and $\mu^i_{(\alpha)}$. At point M, the observers are momentarily colocated and they make a measurement of the same physical quantity. The results of their measurements are related by the transformation between tetrad basis vectors given in Eq. (5.2).

another observer, $\mu^i_{(a)}$, can be contemplated:

$$\lambda^i_{(\alpha)} = H_{(\alpha)}{}^{(\beta)}\mu^i_{(\beta)} \quad \text{for } \alpha, \beta = 1, 2, 3 \tag{5.2}$$

where $H_{(\alpha)}{}^{(\beta)}$ is a 3×3 rotation matrix ($\alpha, \beta = 1, 2, 3$) that relates the spatial tetrad basis vectors. Note that the matrix $H_{(\alpha)}{}^{(\beta)}$ relates three 4-vectors $\lambda^i_{(\alpha)}$ to three 4-vectors $\mu^i_{(\alpha)}$. (For each observer, the 0th components of the tetrad basis, $\lambda^i_{(0)}$ and $\mu^i_{(0)}$, are determined by their respective 4-velocity (see below), so these vectors do not enter the transformation.)

The key idea is that measurements are scalar quantities that are projections on the local basis vectors carried by each observer. Even though two observers coincide in time and space, their tetrad basis vectors are different: $\lambda^i_{(a)}$ for one observer and $\mu^i_{(a)}$ for the other observer. Consequently, the two observers obtain different values of the measurement. Measurements are quantities that are projections on the observer's tetrad; they are of the form

$$F_{ij}\,\lambda^i_{(a)}\lambda^j_{(a)} \tag{5.3}$$

for one observer and of the form

$$F_{ij}\,\mu^i_{(a)}\mu^j_{(a)} \tag{5.4}$$

for the other observer. Each of these quantities is a scalar, i.e., each is invariant under general coordinate transformations given in Eq. (5.1). However, the measured quantities depend on each observer's tetrad and so the measurements are different. At any point, the observer's tetrad is determined by Fermi-Walker transport of the tetrad from some initial point, along the world line of each observer.

The need for the tetrad formalism to relate experiment to theory, as well as the problem of measurable quantities in general relativity, is extensively discussed by Pirani [63], Synge [31], Soffel [64], Brumberg [65], and more recently within the context of metrology by Guinot [66].

The tetrad formalism was initially investigated for the case of inertial observers that move on geodesics [63,67–69,71,74]. Many observers are terrestrially based, or are based on non-inertial platforms and the general theory for the case of non-inertial observers has been investigated by Synge [31], who considered the case of non-rotating observers moving along a time-like world line, and by others [72–78], who considered accelerated, rotating observers. Perhaps the most significant work for space-time navigation in the vicinity of the Earth is contained in Ref. [1,31,71,75,79–81].

As an illustration of the relativistic measurement process, consider an antenna that receives radio frequency electromagnetic waves. The antenna converts an antisymmetric 4-dimensional tensor of second rank, F_{ij}, into a scalar voltage reading on a meter. The meter may have a digital readout of the measurement. Consequently, the voltage is a scalar that does not transform under Lorentz (or generalized coordinate transformations). The voltage that is measured by a moving observer, $V(\tau)$, is a function of the observer's proper time (since some starting point), τ, and depends on the observer's tetrad defined on his world line. The voltage measurement process can be modelled as

$$V(\tau) \quad = \quad F_{ij}M^{ij} \tag{5.5}$$

$$= \quad F_{ij}\lambda^i_{(a)}\lambda^j_{(b)}M^{(ab)} \tag{5.6}$$

$$= \quad F_{(ab)}M^{(ab)} \tag{5.7}$$

where F_{ij} is the electromagnetic field in the space-time, and M^{ij} is the measurement tensor that depends on the world line of the observer. The scalar product of the tensors in Eq. (5.5) reduces the electromagnetic field to a scalar quantity, $V(\tau)$, which is the measured voltage. This voltage $V(\tau)$ is invariant under coordinate transformations of the form in Eq. (5.1). The projection of the electromagnetic field tensor, F_{ij}, on the tetrad basis, $\lambda^i_{(a)}$, yields a set of scalar numbers, $F_{(ab)} = F_{ij}\lambda^i_{(a)}\lambda^j_{(b)}$. The quantity $M^{(ab)}$ is an invariant matrix of numbers that characterizes the measuring apparatus (the observer's antenna). A simple model for an antenna is:

$$M^{(ab)} = \frac{1}{2}\begin{pmatrix} 0 & -l_1 & -l_2 & -l_3 \\ l_1 & 0 & m_3 & -m_2 \\ l_2 & -m_3 & 0 & m_1 \\ l_3 & m_2 & -m_1 & 0 \end{pmatrix} \tag{5.8}$$

where the six quantities, l_α and m_α, represent the sensitivity of the antenna to electric and magnetic fields, respectively. The 3-vector l_α is simply the vector effective length that characterizes the receiving antenna, see Eq. (3.22).

Construction of the Tetrad: Fermi-Walker Transport

The tetrad formalism for a given observer is constructed using the world line of the observer. Consider a time-like world line C of an observer given by coordinates $x^i(u)$ with $u_1 \leq u \leq u_2$, (see Fig. 5.3). Along this world line, the observer has a four velocity $u^i = dx^i/ds$ and acceleration

$$w^i = \frac{\delta u^i}{\delta s} = \frac{du^i}{ds} + \Gamma^i_{jk}u^j u^k \tag{5.9}$$

where Γ^i_{jk} is the affine connection and $\delta u^i/\delta s$ indicates covariant differentiation along the world line $x^i(s)$. The normalization of the 4-velocity, $u^i u_i = -1$, provides the relation between the arc length, $s = c\tau$, where τ is the proper time, and the parameter u.

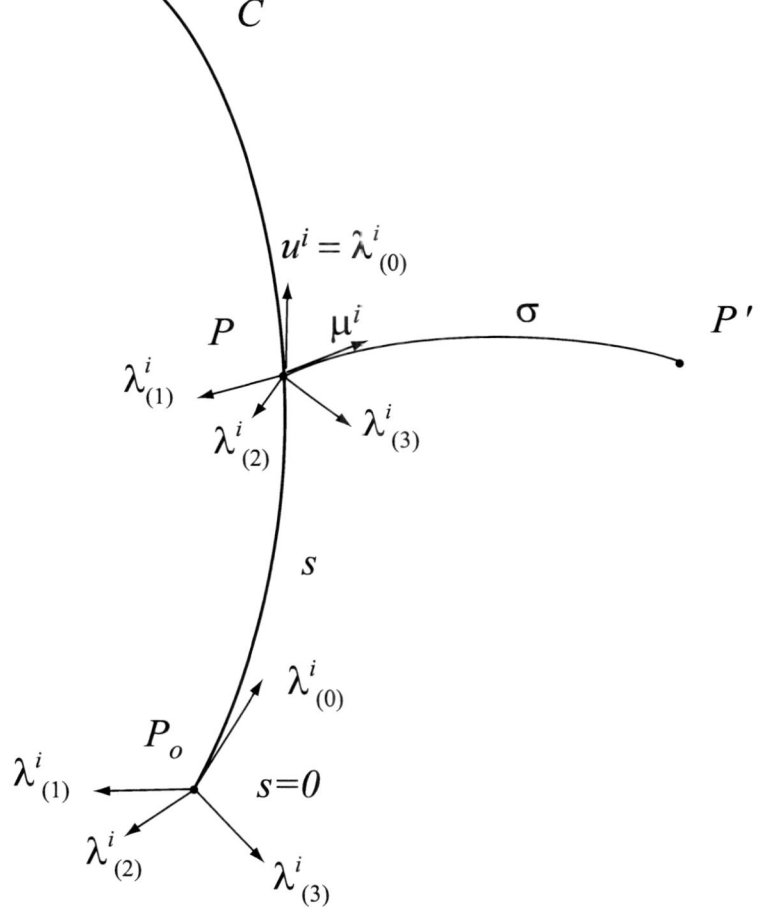

Figure 5.3. The observer's world line C is shown with the initial tetrad basis vectors $\lambda^i_{(\alpha)}$ at $s = 0$ at point P_0. The Fermi transported tetrad basis vectors at proper time s are shown at point P.

As the observer moves on the time-like world line C, he carries with him an ideal clock that keeps proper time τ, and three gyroscopes. At some initial coordinate time x^0, the observer is at point P_0 at proper time $\tau = s/c = 0$. On his world line, the observer carries with him three orthonormal tetrad basis vectors, $\lambda^i_{(\alpha)}$, where $\alpha = 1, 2, 3$, labels the vectors and $i = 0, 1, 2, 3$ labels the components of these vectors in some global system of coordinates. These vectors form the basis for his measurements [31, 63], see Fig. 5.3. The orientation of each basis vector is held fixed with respect to each of the gyroscopes' axis of rotation [82]. The fourth basis vector is taken to be the observer 4-velocity, $\lambda^i_{(0)} = u^i$. The four unit vectors $\lambda^i_{(a)}$, $a = 0, 1, 2, 3$, form the observer's tetrad, which is an orthonormal set of vectors at P_0

$$g_{ij}\,\lambda^i_{(a)}\,\lambda^j_{(b)} = \eta_{ab} \qquad (5.10)$$

where the matrix $\eta_{(ab)}$ is the Minkowski metric, see Appendix B.

At a later time $s = c\tau > 0$, the observer is at a point P. The observer's orthonormal set of basis vectors are related to his tetrad basis at P_0 by Fermi-Walker transport. Fermi-Walker transport preserves the lengths and relative angles of the transported vectors. For an arbitrary vector with contravariant components f^i, its components at P are related to its components at P_0 by the Fermi-Walker transport differential equations [31]

$$\frac{\delta f^i}{\delta s} = W^{ij} f_j \qquad (5.11)$$

where

$$W^{ij} = u^i w^j - w^i u^j \qquad (5.12)$$

When we use Eq. (5.11) to transport a vector f^i that is orthogonal to the 4-velocity, $u^i f_i = 0$, the second term in Eq. (5.12) does not contribute. We refer to transport of such space-like basis vectors as Fermi transport, and $W^{ij} \to \tilde{W}^{ij} = u^i w^j$. The space-like tetrad components satisfy $\lambda^i_{(\alpha)} u_i = 0$, for $\alpha = 1, 2, 3$, and at any point P they are found by integrating the differential Eq. (5.11) over the world line $x^i(s)$, using the initial conditions in Eq. (5.10) on the tetrad at point P_o, with $W^{ij} \to \tilde{W}^{ij}$.

Fermi Coordinates

Associated with each Fermi-transported tetrad basis, there is a set of Fermi coordinates, defined by the geometric construction shown in Figure 5.3. Every event P' in space-time has coordinates $x^{i'}$ in the global coordinate system. According to the observer moving on a time-like world line, the same event has the Fermi coordinates $X^{(a)}$, $a = 0, 1, 2, 3$. The first Fermi coordinate, $X^{(0)} = s$, is taken to be the proper time (in units of length) associated with the event P'. The proper time for P' is defined as the value of arc length s such that a space-like geodesic from point P passes through event P', where the tangent vector of this geodesic, μ^i, is orthogonal to the observer 4-velocity at P:

$$\mu^i u_i|_P = 0 \tag{5.13}$$

The orthogonality condition in Eq. (5.13) is

$$g_{ij}\mu^i(s)\lambda^j_{(0)}(s) = 0 \tag{5.14}$$

and gives an implicit equation for s for a given point P'. This orthogonality condition gives the first Fermi coordinate of the point P'

$$X^{(0)} = s \tag{5.15}$$

The contravariant spatial Fermi coordinates, $X^{(\alpha)}$, $\alpha = 1,2,3$, are defined as [31]

$$X^{(\alpha)} = \sigma\mu^i\lambda^{(\alpha)}_i = g_{ij}\sigma(s)\mu^i(s)\eta^{(\alpha\beta)}\lambda^j_{(\beta)}(s) \tag{5.16}$$

where σ is the measure along the space-like geodesic between P and P', g_{ij} is the metric and $\eta^{(ij)} = \eta_{(ij)} = \eta_{ij}$ is the invariant Minkowski matrix, see Appendix B.

Metric in Fermi Coordinates

All measurements made by a real observer are done locally, at the origin of Fermi coordinates. The measurements are projections on the tetrad of the observer, and the tetrad is only defined on the world line of the observer. Never-the-less, Fermi coordinates can be defined off the world line of the observer, and a corresponding metric for Fermi coordinates can be defined. The space-time interval in the Fermi coordinate system of the observer is

$$ds^2 = -G_{(ij)}\,dX^{(i)}\,dX^{(j)} \tag{5.17}$$

where $G_{(ij)}$ are the metric tensor components when the $X^{(i)}$ are used as coordinates [31].

Despite the clear interpretation of measurement that the tetrad formalism offers, the analysis of experiments has seldom been done using the full tetrad formalism described above. In part, this is due to limitations of data accuracy and theorist patience to carry out the detailed computations. There are, however, a few explicit theoretical constructions of tetrads in the literature that address the issues discussed here [71, 75, 78, 80, 81].

Chapter 6

Reference Frames and Coordinate Systems

The world function of space-time (see Section V) is a useful tool for describing the problem of navigation in space-time in a simple geometric way. The general equations are covariant, and invariant, and there is no need to specify a system of space-time coordinates. Navigation is carried out by making local measurements, in the comoving frame of the apparatus that does the measurement.

However, in actual applications, a specific implementation of a system of space-time coordinates must be used. In most real-life applications today, such as the GPS, a fully relativistic 4-dimensional space-time coordinate system is not used. Instead, a system of three-dimensional coordinates plus a set of clocks are used. There are three common systems of three-dimensional coordinates that play a role in the navigation problem: Earth-centered inertial (ECI) coordinates, Earth-centered Earth-fixed (ECEF) coordinates, and topocentric coordinates. The origin of ECI coordinates is at the Earth's center of mass, and the orientation is determined with respect to distant objects–so the coordinates do not rotate with respect to these distant objects. However, the origin of ECI coordinates revolves around the sun along with the Earth, so these coordinates are better called quasi-inertial coordinates. ECEF coordinates have the same origin as ECI coordinates, but these coordinates rotate with the Earth, so a point that is stationary on the Earth surface has a constant value for its ECEF coordinates. Topocentric coordinates have their origin on the Earth surface, with the x-axis pointing South, the y-axis pointing East, and the z-axis pointing radially away from Earth center (pointing up). Topocentric coordinates are used for making radar and other observations on the Earth surface. All three of these coordinate systems are three-dimensional. In experimental observations, these three dimensional coordinate systems are used together with clocks to record the coordinates of space-time events. Great care must be exercised when relativistic theories are used to analyze the data, because the observations were not recorded with respect to a true (relativistic) four-dimensional system of space-time coordinates. As an example of potential errors, see the next section.

6.1. Gravitational Warping of Coordinates

The presence of a gravitational field leads to a warping of the system of coordinates. More precisely, if we attempt to use Euclidean space and time coordinates, and neglect the effect of a gravitational field on the coordinate system, we will find that the definition of these coordinates is not precise. In other words, the definition of these Euclidean coordinates is ill-defined when a gravitational field is present. To demonstrate the magnitude this effect, consider the Schwarzschild metric

$$-ds^2 = g_{ij}dx^i dx^j = (1 - \frac{r_g}{r})c^2 dt^2 - \frac{dr^2}{1 - \frac{r_g}{r}} - r^2(\sin^2\theta d\phi^2 + d\theta^2) \qquad (6.1)$$

where the gravitational radius is given by $r_g = 2GM/c^2$. For Earth, $M = 5.98 \times 10^{24}$ kg, so $r_g = 0.88$ cm. The physical length dl in a static space-time is given by [32]

$$dl^2 = \gamma_{\alpha\beta}\, dx^\alpha dx^\beta = \frac{dr^2}{1 - \frac{r_g}{r}} + r^2(\sin^2\theta d\phi^2 + d\theta^2) \qquad (6.2)$$

where the 3-dimensional spatial metric $\gamma_{\alpha\beta}$ is related to the 4-dimensional metric g_{ij} by

$$\gamma_{\alpha\beta} = g_{\alpha\beta} - \frac{g_{0\alpha}g_{0\beta}}{g_{00}} \qquad (6.3)$$

The physical distance between two points at radial coordinates r_1 and r_2, and at the same coordinate angle θ and ϕ, is given by

$$l = \int_{r_1}^{r_2} dl = \int_{r_1}^{r_2} \frac{dr}{\sqrt{1 - \frac{r_g}{r}}} = \left[\frac{(r - r_g)}{\sqrt{1 - \frac{r_g}{r}}} + r_g \log(\sqrt{r} + \sqrt{r - r_g}) \right]_{r_1}^{r_2} \qquad (6.4)$$

Now imagine the first point is on the Earth's equator, so that $r_1 = 6.378 \times 10^6$ m is the Earth's equatorial radius, and the second point is at the altitude of a GPS satellite where $r_2 = 2.6561 \times 10^7$ m. Since the Earth's gravitational radius is $r_g = 2GM/c^2 = 0.88$ cm, the length l in Eq. (6.4) can be expanded in the small parameter $x = r_g/r_1 << 1$, giving

$$l = r_2 - r_1 + \frac{1}{2} r_g \log \frac{r_2}{r_1} \qquad (6.5)$$

The first term on the right, $r_2 - r_1$, is the Euclidean length between the coordinate points at radial coordinate r_1 and at r_2. The second term is the correction to the physical length l due to the presence of a gravitational field. The magnitude of this correction, for the r_1 and r_2 values cited above, is $\frac{1}{2} r_g \log \frac{r_2}{r_1} = 0.63$ cm. This shows that the physical length l is longer by 0.63 cm than the difference between radial coordinate values in a Euclidean space that has zero gravitational field. The gravitational field has the effect of stretching the physical space between coordinate values.

The implication of the effect is that a 4-dimensional relativistic frame of reference (coordinate system) must be implemented (rather than a system of 3-dimensional coordinates plus a time scale) if we want to have an unambiguous definition of a space-time coordinate system. In other words, a Euclidean system of 3-dimensional coordinates, plus a time scale, will have inherent error of $\frac{1}{2} r_g \log \frac{r_2}{r_1}$ over a distance $r_2 - r_1$, due to the warping of space-time due to the presence of the gravitational field.

In the above calculation, I have neglected the Earth's quadrupole moment $J_2 = 1.0826800 \times 10^{-3}$ of the mass distribution. The inclusion of this quadrupole moment can be expected to modify the gravitational correction (second term) in Eq. (6.5) by additional terms of order $J_2 r_g$, leading to an additional length correction on the order of $J_2 r_g \log \frac{r_2}{r_1}$, which is of the order of 10^{-2} mm over distances $r_2 - r_1 = 2.0 \times 10^7$ m.

Another way of stating the effect of the gravitational warping of coordinates is to say that, if we used Euclidean geometry, the accuracy of clock synchronization is limited to $(0.63\text{cm})/c = 2.1 \times 10^{-11}$ s $= 21$ ps, even if perfect clocks and equipment are used. This value of 21 ps assumes the previous values of r_1 and r_2. Reference to Table I shows that, for some applications, we must have time synchronization to better than 21 ps.

The gravitational warping of coordinates is closely associated with (but distinct from) the slowing down of light in a gravitational field, also known as the

Shapiro time delay effect. The average speed of light can be computed along a radial path from r_1 to r_2. The path is a null geodesic given by $ds^2 = 0$, which gives $c^2 dt^2 = dr^2/(1 - r_g/r)^2$. The coordinate time for light to traverse this path is

$$\int_{t_1}^{t_2} dt = \frac{1}{c} \int_{r_1}^{r_2} \frac{dr}{1 - \frac{r_g}{r}} \tag{6.6}$$

The average speed of light over this path is then given by

$$\frac{l}{t_2 - t_1} = c \left[1 - \frac{1}{2} \frac{r_g}{r_2 - r_1} \log \frac{r_2}{r_1} \right] \tag{6.7}$$

The second term in the above equation is the correction to the speed of light due to the presence of the Earth's mass, and for the previous values of r_1 and r_2 has the small magnitude

$$\frac{\Delta c}{c} = -\frac{1}{2} \frac{r_g}{r_2 - r_1} \log \frac{r_2}{r_1} = -3.13 \times 10^{-10} \tag{6.8}$$

The negative sign in Eq. (6.7) means that light is has travelled slower than in vacuum in the absence of a gravitational field. This correction to the speed of light depends on the direction of light travel.

Chapter 7

Clock Synchronization

As discussed in the introduction, accurate clock synchronization is the backbone of applications such as high-accuracy navigation, communication, geolocation, and space-based interferometer systems. Also, as previously discussed, clock synchronization cannot be divorced from the general problem of navigation in space-time (determining the position and time of an observer). The key point here, is that clock synchronization (divorced from the problem of determining spatial coordinates) is not a covariant concept, whereas navigation in space-time *is* a covariant concept, and consequently it can be formulated in covariant equations, such as was done in Subsection V-A, using the world function. See Eq. (4.6). However, if there exists a system of coordinates in which the clocks are all stationary at constant known spatial positions, then we can discuss their synchronization. However, this is generally not the case because satellites that carry clocks orbit the Earth.

In the next Subsection B, we discuss synchronization of clocks that are stationary, at known spatial positions of an arbitrary system of coordinates, in the presence of a gravitational field. The system of coordinates need not be rigid and can be rotating. Section C discusses the peculiar way that GPS clocks are "synchronized". Since the GPS satellites are moving, the clocks are (approximately) locally synchronized to a coordinate time in a certain metric, and hence the meaning of synchronization is different for GPS satellites than for stationary clocks.

In pre-relativistic physics, clock synchronization was a straight-forward concept wherein two clocks were simply set to the same time. With the advent of highly accurate atomic clocks on board satellites, more accurate schemes for synchronizing remote clocks have been developed. Furthermore, the clocks to be synchronized are in relative motion as well as in different gravitational potentials. The accuracy of clocks has improved to the point that previously small unobservable effects must be accounted for by a comprehensive theory. A metric theory of gravity, such as general relativity, is the vehicle of choice for dealing with clock synchronization. Within the theory of relativity there are a number of clock synchronization schemes. However, one central concept–that of simultaneous event–is at the heart of the issue of clock synchronization. Simultaneity of two events is a definition, and not an absolute covariant concept, unless the two events are co-located at the same event in space-time.

7.1. Eddington Slow Clock Transport

Clocks on maritime vessels of the 1700 and 1800's were used to navigate the oceans. An accurate clock was kept on the ship, and it was used to determine longitude of the vessel's current position [83]. The basic idea was to carry an accurate clock, and be in possession of the "correct time". Using this time, and a view of the sky, the ship's longitude could be computed based on a theory of Earth rotation. This scheme of having the correct time by slowly transporting a clock is often called "Eddington slow clock transport" [84].

From the point of view of special relativity, where $g_{00} = -1$ and $g_{\alpha\beta} = \delta_{\alpha\beta}$, $\alpha, \beta = 1, 2, 3$. (constant and uniform gravitational field), a clock can be transported with arbitrarily small velocity and still maintain the "correct time". From Eq. (2.2), when the velocity of a clock is arbitrarily small, $dx^\alpha/dx^0 \to 0$, $\alpha = 1, 2, 3$, the relation between proper time τ and coordinate time $t = x^0/c$, reduces to $\Delta\tau = \Delta x^0/c$. So a slow moving clock, which keeps proper time τ, can by definition be made to keep coordinate time x^0. Such a clock can be slowly transported over large distances and can be used to synchronize other remote clocks. Unfortunately, Eddington slow clock transport is too restrictive for applications to clocks on satellites, because the speed of the satellites is not small, typically on the order of $v/c = dx^\alpha/dx^0 \sim 10^{-5}$, and because the gravitational potential between different satellites varies significantly. See Section IX for a discussion of the effect of satellite motion and gravitational potential on proper time.

7.2.   Einstein Synchronization

Another method of synchronizing *stationary* clocks is based on exchanging electromagnetic signals between clocks at *known spatial locations*. This type of synchronization is often called Einstein synchronization, and is based on a particular definition of simultaneity of (spatially separated) events. Perhaps the clearest discussion of the definition of simultaneity, within the context of general relativity, is given by L. D. Landau and E. M. Lifshitz [32]. We use their definition of simultaneity in the clock synchronization argument presented below.

Consider two clocks at rest in some frame of reference (4-dimensional coordinates). Clock A has world line defined by constant spatial coordinates x^α, and clock B has world line defined by spatial coordinates $x^\alpha + dx^\alpha$, $\alpha = 1, 2, 3$, see Figure 7.1. Clock B is to be synchronized to clock A by exchange of electromagnetic signals. At event P_1, with coordinate time $x^0 + dx_1^0$, an electromagnetic signal is sent from clock B to clock A. Clock A receives the signal at event P_4, which has coordinates (x^0, x^α), and immediately reflects the signal back to clock B, where it is received at event P_2 with coordinates $(x^0 + dx_2^0, x^\alpha + dx^\alpha)$. This reflected signal (which is received by B at P_2) contains the proper time reading, τ_A, of clock A at event P_4.

The question then arises: what point P (coordinate time) on world line of clock B is simultaneous with event P_4 on world line of clock A? Detailed consideration of this problem leads to the conclusion that there is no preferred way to define a point on world line B that is simultaneous with event P_4. Therefore, we *arbitrarily* take the midpoint between P_1 and P_2 as defining the point P that is simultaneous with P_4. For electromagnetic propagation, the space-time interval must vanish:

$$-ds^2 = g_{ij}dx^i dx^j \tag{7.1}$$
$$= g_{\alpha\beta}dx^\alpha dx^\beta + 2g_{0\alpha}dx^0 dx^\alpha + g_{00}\left(dx^0\right)^2 = 0$$

Solving for the coordinate time dx^0, we get two solutions, corresponding to the two directions of signal travel between x^α and $x^\alpha + dx^\alpha$:

$$dx_1^0 = -\frac{1}{g_{00}}\left(g_{0\alpha}dx^\alpha - \sqrt{(g_{0\alpha}g_{0\beta} - g_{00}g_{\alpha\beta})dx^\alpha dx^\beta}\right) \tag{7.2}$$

$$dx_2^0 = -\frac{1}{g_{00}}\left(g_{0\alpha}dx^\alpha + \sqrt{(g_{0\alpha}g_{0\beta} - g_{00}g_{\alpha\beta})dx^\alpha dx^\beta}\right) \tag{7.3}$$

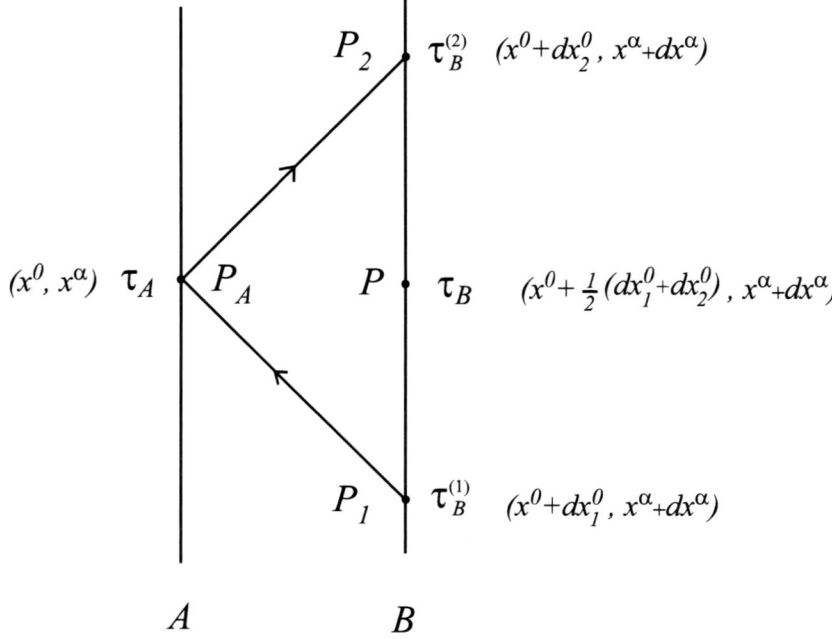

Figure 7.1. The world line of clock A and clock B are shown. Clock B is synchronized to clock A by electromagnetic signals. A signal from clock B is sent out at event P_1, the signal reflects off the face of clock A at event P_A, and returns to clock B at event P_2.

Note that if $dx_2^0 > 0$ then $dx_1^0 < 0$. Therefore, the coordinate time of event P, which is simultaneous with event P_A, is given by (using the midpoint definition of simultaneity):

$$x^0 - \frac{g_{0\alpha}dx^\alpha}{g_{00}} \qquad (7.4)$$

Since events P and P_A are simultaneous (by definition), we consider clock A and B to be synchronized when their proper times are equal at these events:

$$\tau_B = \tau_A \qquad (7.5)$$

This corresponds to shifting the origin (epoch) of proper time.

The reflected signal from clock A arrives at B at the time $\tau_B^{(2)}$. The proper time interval between event P and P_2 is given in terms of coordinate time by the integral:

$$\tau_B^{(2)} - \tau_B = \frac{1}{c} \int_P^{P_2} ds \tag{7.6}$$

Now, consider clocks A and B to be stationary, so that their spatial coordinates, x^α and $x^\alpha + dx^\alpha$, are constant in time in the chosen coordinate system. Then the proper time elapsed on clock A between point P and P_A is given by

$$\tau_B^{(2)} - \tau_B = \frac{1}{c} \int_{x^0 + \frac{1}{2}(dx_1^0 + dx_2^0)}^{x^0 + dx_2^0} \sqrt{-g_{00}} \, dx^0 \tag{7.7}$$

$$= \frac{1}{c} \left[\frac{1}{\sqrt{-g_{00}}} \sqrt{(g_{0\alpha}g_{0\beta} - g_{00}g_{\alpha\beta})dx^\alpha dx^\beta} \right]_P \tag{7.8}$$

where in the last line the quantities are evaluated at P, at coordinate time given by Eq. (7.4). Using the condition in Eq. (7.5), Eq. (7.8) can be written as

$$\tau_B^{(2)} = \tau_A + \frac{1}{c} \left[\frac{1}{\sqrt{-g_{00}}} \sqrt{(g_{0\alpha}g_{0\beta} - g_{00}g_{\alpha\beta})dx^\alpha dx^\beta} \right]_P \tag{7.9}$$

Equation (7.9) gives the condition for clock B to be synchronized to clock A, in terms of quantities that are observed by clock B. Note that τ_A is the proper time on clock A, as observed by clock B. Equation (7.9) gives the proper time that must be set on clock B at event F_2 so that clock B is "synchronized" with clock A at the earlier event P. Note that the synchronization was done between clocks A and B that are separated by an infinitesimal spatial interval dx^α. In practice, for small gravitational fields, this infinitesimal interval can correspond to large distances. For the case of a flat space-time with a Minkowski metric, $g_{00} = -1, g_{\alpha\beta} = \delta_{\alpha\beta}$, Eq. (7.9) gives $\tau_B^{(2)} = \tau_A + l/c$ where the known spatial distance between clocks is $l = g_{\alpha\beta}dx^\alpha dx^\beta$.

Some comments are in order on the practical application of the synchronization condition in Eq. (7.9). From Eq. (7.4), we know the space-time coordinates of point $P = (x^0 - \frac{g_{0\alpha}dx^\alpha}{g_{00}}, x^\alpha + dx^\alpha)$, and the synchronization condition in Eq. (7.9) depends on the metric components g_{ij}. For high-accuracy clock synchronization, we must know the metric components g_{ij} in the coordinate system in which the clocks are at rest. For the accuracy needed in some applications (see Table 1.1), the metric is not known to sufficient accuracy. An even more serious problem is that the spatial positions of clock A and B must be known a priori, i.e., the clock positions are not determined by the synchronization protocol, and the synchronization protocol depends on these positions. Finally, in the real world, satellites are in motion with respect to one another and there is no single (simple) coordinate system in which more than two satellites are at rest.

In conclusion, it becomes evident that for most applications, the critical problem is not clock synchronization (correlating clock times), but that of navigation (correlation of clock positions and times). See Section V subsection A.

Consequently, in real applications such as in the satellite system known as GPS, a different synchronization scheme is used. In the GPS scheme, satellite clocks keep (approximately) the coordinate time in the underlying ECI coordinate frame. We discuss GPS clock synchronization in the next section.

In practice, Einstein clock synchronization is the basis of the applied technique known as two-way satellite time transfer (TWSTT) which in practice gives very accurate time synchronization between two points that are in common view of the same communication satellite [89, 90].

7.3.   GPS Clock Synchronization

The Global Positioning System (GPS) is a U.S. Department of Defense Satellite System consisting of approximately 24 satellites orbiting the Earth. The system consists of four satellites in each of six inclined orbital planes of $55°$. Each satellite of the GPS carries atomic clocks, and sends pseudorandom code signals that are the basis of providing accurate time and navigation signals [2, 85–87]. The GPS uses a unique time synchronization scheme, wherein the clocks send time, encoded on chips of the pseudorandom code, and these chips (with time stamps on them) are received at Earth monitoring stations. The stations determine any clock corrections needed based on the broadcasts from the satellites. The GPS satellites move at a fast speed (approximately $v/c \sim 10^{-5}$) and the clocks suffer relativistic time dilation of approximately $7\,\mu s$ per day. On the other hand, the satellites are at a high altitude, and the clocks run fast (as compared with coordinate time in the ECI frame metric), by about $45\,\mu s$ per day, due to the high gravitational potential in which they operate [2]. The net effect is that, with respect to a clock on the Earth's surface, the clocks appear to run fast by approximately $38\,\mu s$ per day.

Each GPS satellite has an orbit that is approximately circular. Consequently, all the clocks on the satellites behave in roughly the same way–they run fast by $38\,\mu s$ per day, or in terms of frequency, as observed from the surface of the Earth

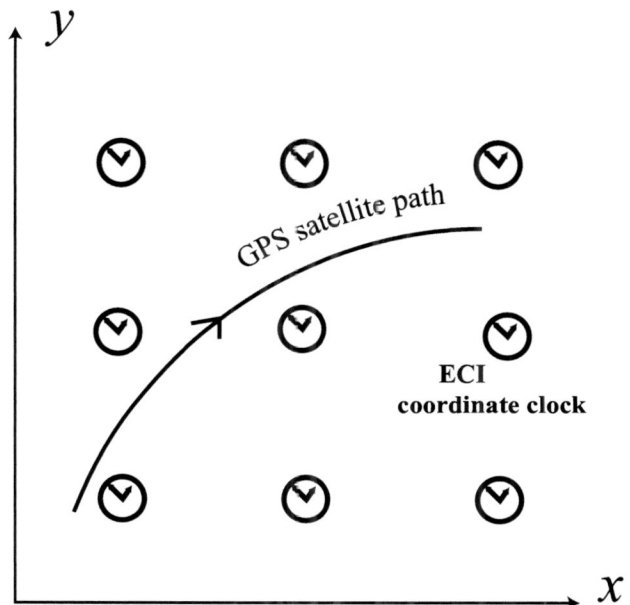

Figure 7.2. The motion of a GPS satellite in 3-dimensional space is shown as it passes (stationary) coordinate clocks in the ECI frame. At any instant, the GPS satellite clock agrees with time on the underlying ECI frame coordinate clocks.

the oscillators run fast by 1 part in 4.4×10^{-10}, see Section IX, subsection E. If something were not done to counter this effect, all GPS satellite clocks would appear to run fast. Consequently, a "factory offset" is applied to the frequency of each clock oscillator (in software) in the amount -4.4×10^{-10}. The clocks on board the GPS satellites then keep (approximately) coordinate time in the ECI frame, and the clocks appear to (approximately) keep correct time as seen from the Earth's surface. This "factory offset" does not compensate for the different orbital eccentricities of the satellites, since for each satellite. The factory also does not compensate for the relative velocity between satellites and different moving GPS receivers, see Ref. [2] for a detailed discussion of the role of the Doppler effect in GPS measurements.

The "factory offset" applied to the GPS satellite oscillators can be understood as another scheme for clock synchronization. There exists a coordinate time $t = x^0/c$, in some metric around the Earth, see Eq. (8.1). The coordinate time is a global quantity, as compared to the proper time, which depends on the given world line. By definition, clocks on the Earth's geoid keep coordinate time in the ECEF frame. In order to synchronize the GPS satellite clocks to coordinate time in the ECI frame, the "factory offset" is applied. With this offset, the GPS clocks keep approximate coordinate time in the ECI frame of reference. As each satellite moves through space, we can imagine that it passes "hypothetical" coordinate clocks (that are stationary in the ECI frame and keep ECI coordinate time). The "factory offset" has the effect that a GPS satellite clock instantaneously agrees (closely) with each ECI frame coordinate clock that it passes, see Figure 7.2. So the GPS satellite clocks are (approximately) synchronized to coordinate clocks in the underlying ECI frame. Actually, this is an unusual idea–for (GPS) clocks in one frame (comoving frame of a GPS satellite) to keep coordinate time in another (ECI) frame. As mentioned above, in the case of GPS, this synchronization is approximate, because only a constant rate offset is applied before satellite launch, which cannot compensate for effects, such as the eccentricity of each satellite orbit (called $e \sin E$ effect), the Earth's quadrupole moment (or J_2 effect), the effects of other forces (such as solar pressure and atmospheric drag) on each satellite, and imperfect atomic clocks. The satellite ground tracking system for the GPS monitors satellite clocks and attempts to compensate for these effects by calculating satellite clock corrections and uploading these corrections to the satellites so that users of the GPS may apply them to get accurate space-time navigation.

7.4. Quantum Synchronization

During the past several years, alternative schemes for clock synchronization [21–30] have been proposed based on quantum information theory and entanglement of quantum states [19, 20]. Quantum effects may be exploited for clock synchronization since entangled (or correlated) photon pairs are found to interfere destructively at a beam splitter [21]. Such photon pairs are believed to be created at a single space-time event, within 100 fs of each other [21]. Exploiting this effect, two relatively simple synchronization schemes have been very recently proposed based on the interference of entangled photon pairs states, which can be created by parametric down conversion in a crystal that lacks a center of inversion symmetry [21]. One of these schemes is capable of clock synchronization in free space [29], while the other relies on there being a difference of group velocities for each of the photons in the entangled pair in an optical medium, such as exists in an optical fiber [30]. Very recently, an experimental demonstration of quantum clock synchronization has been carried out in the laboratory [30].

The quantum mechanical clock synchronization proposals do not included the basic relativistic effects present when clocks on board satellites are to be synchronized: the fast relative motion of clocks and the variation in gravitational potential between satellites in various orbital regimes, such as low Earth orbit (LEO), geosynchronous orbit (GEO), and highly elliptical orbit (HEO). In the next section, we describe the resulting syntonization problem that arises for satellites at differing gravitational potentials.

Chapter 8

The Syntonization Problem

Satellites in multiple orbital regimes may have their clocks synchronized by means of exchange of optical signals, as described in the previous section. This means that at one coordinate time, all clocks can be made to read the same value (for example, the same coordinate time). Satellite applications require that the synchronization between members of the satellite ensemble be maintained for some time period, or alternatively, the clock differences must be known at a given elapsed time from the synchronization epoch. It is well known that the proper time, and consequently the hardware time (see Section III), will run at a different rate on each satellite with respect to coordinate time of some given metric. This difference in rate of proper time is due to the motion of the clock (time dilation) and gravitational potential effects (sometimes called the gravitational red shift effect).

Since the hardware time on all real clocks is affected by their motion and position in the space (local gravitational potential), the only reasonable measure of time is coordinate time, as used in a metric theory of gravity, such as general relativity. Coordinate time is a mathematical construct that is global, which means that it is the same everywhere in the space. In distinction, proper time depends on the history (world line) of the clock. In this Section, we compute the difference between proper time on board a satellite, and coordinate time, for satellites in various orbits about the Earth.

Since the clock on board a satellite is located at a different spatial position than a reference clock on the Earth's surface, these clocks can be compared by exchange of electromagnetic signals. The relation between these two clocks must be established, such that their relative motion and their respective gravitational potentials are taken into account. At any instant in time, the relative velocity of the satellite with respect to the ground leads to the satellite clock running slower than the ground clock, due to relativistic time dilation. The higher gravitational potential at the satellite leads to the satellite clock running faster than the ground clock. Generally speaking, clocks in low-orbiting satellites run slow due to the predominant time dilation effect, due to high orbital velocity and small altitude above the Earth's surface. On the other hand, clocks in high-orbit satellites generally run faster than ground clocks because the gravitational potential effect is predominant.

While a satellite clock can be compared to an Earth-bound reference clock, there is nothing special about the Earth-bound clock as a reference clock. In fact, from the point of view of an ECI coordinate system, the Earth bound clock moves in a circle, just like the satellite clock. Furthermore, the actual reading on an ideal clock depends on its world line, i.e., its past history of velocity and gravitational potential. All ideal (and hardware) clocks suffer from this complication. Therefore, the comparison of all clocks must be made to a standard time or "clock" that is global and does not depend on its history. Such a quantity is the coordinate time associated with some metric that describes the space-time in the vicinity of the Earth. Coordinate time is a mathematical construct that is the same in all of space-time. The complication is that coordinate time (and generally each spatial coordinate) is arbitrary in a metric theory such as general relativity, and depends on an arbitrary choice of a space-time coordinates. In the next section, we will choose a space-time metric that has the desirable property that coordinate time corresponds (approximately) to the proper time kept by an ideal clock on the Earth's geoid (surface of equal geopotential).

8.1. Choice of Metric in Vicinity of the Earth

The definition of coordinate time comes from choosing a specific metric for the space-time. In general relativity, the coordinates x^i, $i = 0, 1, 2, 3$, are mathematical entities that are never observed. The metric of space-time depends on these coordinates, and takes into account the warping of space-time due to the presence of the gravitational field. For the same physical space-time, we can choose different coordinates. Of course, all observations are independent of the coordinates, and so the choice of coordinates is arbitrary. However, it is convenient to choose space-time coordinates so that coordinate time has some physical meaning. Perhaps the most reasonable choice is to take the metric for space-time of the form [88]

$$-ds^2 = -(1+\frac{2}{c^2}V)(d\bar{x}^0)^2 + (1-\frac{2}{c^2}V)\left[(dx^1)^2 + (dx^2)^2 + (dx^3)^2\right] \quad (8.1)$$

where $\bar{x}^0, x^1, x^2, x^3$ are geocentric equatorial coordinates, where x^3 coincides with the Earth's axis of rotation and increasing positive values point to North. The Earth is taken to be an oblate spheroid with potential given by [91]

$$V(r,\theta) = -\frac{GM}{r}\left[1 - J_2\left(\frac{R_e}{r}\right)^2 P_2(\cos(\theta))\right] \quad (8.2)$$

where, $r^2 = (x^1)^2 + (x^2)^2 + (x^3)^2$ and θ is the polar angle measured from the x^3 axis. In Eq. (8.2), G is Newton's gravitational constant, M is the mass of the earth, $P_2(x) = (3x^2 - 1)/2$ is the second Legendre polynomial, R_e is the Earth's equatorial radius, and J_2 is the Earth's quadrupole moment, whose value is approximately $J_2 = 1.0826800 \times 10^{-3}$. The metric in Eq. (8.1) is the solution [32] of the linearized version of Einstein Eqs. (3.1).

The coordinate time can be given a simple interpretation. Transform the metric in Eq. (8.1) to rotating ECEF coordinates y^i, using the transformation

$$\begin{aligned}
\bar{x}^0 &= \bar{y}^0 \\
x^1 &= \cos(\frac{\omega}{c}\bar{y}^0)y^1 - \sin(\frac{\omega}{c}\bar{y}^0)y^2 \\
x^2 &= \sin(\frac{\omega}{c}\bar{y}^0)y^1 + \cos(\frac{\omega}{c}\bar{y}^0)y^2 \\
x^3 &= y^3
\end{aligned} \quad (8.3)$$

Note that the coordinate time in the rotating frame, $\bar{y}^0 = \bar{x}^0$. In these ECEF rotating coordinates, the metric is given by

$$\begin{aligned}
-ds^2 &= -\left[1 + \frac{2V}{c^2} - \frac{\Omega^2}{c^2}\left[(y^1)^2 + (y^2)^2\right] + \frac{2V}{c^2}\frac{\Omega^2}{c^2}\left[(y^1)^2 + (y^2)^2\right]\right](d\bar{y}^0)^2 \\
&\quad + (1 - \frac{2V}{c^2})\left[2\frac{\Omega}{c}(y^1 dy^2 - y^2 dy^1)d\bar{y}^0 + (dy^1)^2 + (dy^2)^2 + (dx^3)^2\right] (8.4)
\end{aligned}$$

From Eq. (8.4), we see that stationary clocks in the ECEF coordinates (clocks which satisfy $dy^\alpha = 0$) that are at the same value of geopotential ϕ,

where

$$\phi = V - \frac{1}{2}\Omega^2 \left((y^1)^2 + (y^2)^2\right) \tag{8.5}$$

have the same rate of proper time with respect to coordinate time y^0. The term geopotential is used because ϕ takes into account the effect of angular velocity Ω of Earth rotation. We have neglected the small cross-term $\frac{2V}{c^2}\frac{\Omega^2}{c^2}R^2 \sim 10^{-21}$.

Using the observation that clocks at a constant value of geopotential run at the same rate, it is advantageous to define the new coordinate time t

$$ct = x^0 = y^0 = \left(1 + \frac{\phi_o}{c^2}\right)\bar{y}^0 = \left(1 + \frac{\phi_o}{c^2}\right)\bar{x}^0 \tag{8.6}$$

where ϕ_o is the geopotential on the Earth's equator:

$$\phi_o = -\frac{GM}{R_e}(1 + \frac{1}{2}J_2) - \frac{1}{2}\Omega^2 R_e^2 \tag{8.7}$$

The dimensionless magnitude of this term is $\phi_o/c^2 = -6.96928 \times 10^{-10}$. Using the transformation in Eq. (8.6), the metric becomes

$$
\begin{aligned}
-ds^2 &= -\left[1 + \frac{2}{c^2}(\phi - \phi_o)\right](dy^0)^2 + \left(1 - \frac{2V + \phi_o}{c^2}\right)2\frac{\Omega}{c}(y^1 dy^2 - y^2 dy^1)dy^0 \\
&\quad + (1 - \frac{2V}{c^2})\left[(dy^1)^2 + (dy^2)^2 + (dx^3)^2\right]
\end{aligned} \tag{8.8}
$$

Eq. (8.8) gives the space-time metric in ECEF rotating coordinates y^i. Note that an ideal clock that is stationary in EFEC coordinates (with $dy^\alpha = 0$), has proper time

$$d\tau = ds/c = \frac{1}{c}\left[1 + \frac{2}{c^2}(\phi - \phi_o)\right]^{1/2} dy^0 \tag{8.9}$$

When this clock is located on the geoid, then $\phi = \phi_o$, and $d\tau = dy^0/c$, so this ideal clock keeps coordinate time. Hence a good hardware clock on the geoid can be used as a standard to keep coordinate time, x^0, in the rotating ECEF coordinates. Note that by Eq.(8.3) the coordinate time in rotating ECEF coordinates is the same as coordinate time in ECI coordinates, so this same clock keeps coordinate time, x^0, in the ECI frame coordinates x^i.

Using the coordinate time transformation in Eq. (8.6), the metric in the ECI coordinates, given in Eq. (8.1), becomes

$$-ds^2 = g_{ij}dx^i dx^j =$$
$$-\left[1 + \frac{2}{c^2}(V - \phi_o)\right](dx^0)^2 + \left(1 - \frac{2}{c^2}V\right)\left[(dx^1)^2 + (dx^2)^2 + (dx^3)^2\right] \quad (8.10)$$

Equation (8.10) gives the metric in ECI coordinates. The coordinate time that enters into the metric, x^0, is the time kept by ideal clocks on the geoid. This result was the singular goal of the time transformation given in Eq. (8.6).

Note however, that in the ECI metric in Eq. (8.10), the proper time interval ds on a stationary clock in ECI coordinates (with $dx^\alpha = 0$), is not equal to coordinate time interval dx^0 because in general $V \neq \phi_o$.

The ECI coordinate metric, given in Eq. (8.10), will be used below as the basis for comparing satellite clocks in different orbits. The time kept by ideal satellite clocks (proper time) will be compared to the global coordinate time x^0, which is the time kept by ideal clocks on the geoid.

8.2.   Integration of Geodesic Equations

In order to compare the proper time on a satellite clock to the coordinate time in ECI metric given in Eq. (8.10), the satellite world line must be known. The satellites mass is negligible compared to that of the Earth, so the satellite is essentially a freely falling test particle that follows a geodesic in 4-dimensional space-time. The geodesic equations are given by

$$\frac{d^2 x^i}{ds^2} + \Gamma^i_{jk}\frac{dx^j}{ds}\frac{dx^k}{ds} = 0 \quad (8.11)$$

where the coordinates are given by $x^i = (x^0, x^\alpha)$, $\alpha = 1, 2, 3$. The four velocity can be written as

$$v^i(s) = \frac{dx^i(s)}{ds} = \frac{dx^0}{ds}\left(1, \frac{dx^\alpha}{dx^0}\right) \equiv \frac{dx^0}{ds}v^i \quad (8.12)$$

where $v^i = (1, dx^\alpha/dx^0) = (1, dx^\alpha/d(ct))$. The 4-velocity is normalized,

$$-1 = \left(\frac{dx^0}{ds}\right)^2\left[-\alpha - \beta\delta_{\kappa\gamma}v^\kappa v^\gamma\right] \quad (8.13)$$

where α and β are functions of s.

The four geodesic differential Eq. (8.11) must be supplemented by initial conditions: initial 4-velocity $v^i(s=0)$ and an initial 4-dimensional position $x^i(s=0)$. These initial conditions are taken from the classical Newtonian orbital mechanics in 3-dimensional Euclidean space.

Now introduce Cartesian perifocal coordinates (z^0, z^1, z^2, z^3), defined so the origin coincides with one focus and the $+z^1$-axis contains the periapsis, and the z^1-z^2 plane contains the elliptical orbit [92]. In these perifocal coordinates z^i, the Cartesian coordinate position $\mathbf{r}$ and velocity $\mathbf{v}$ are given by

$$\mathbf{r} = r\cos v\mathbf{P} + r\sin v\mathbf{Q} \tag{8.14}$$

$$\mathbf{v} = \left(\frac{GM}{a(1-e^2)}\right)^{1/2}[-\sin v\mathbf{P} + (e+\cos v)\mathbf{Q}] \tag{8.15}$$

where v is the true anomaly, a is the semimajor axis of the orbit, and e is the orbital eccentricity. In Eq. (8.14)–(8.15), $\mathbf{P}$ and $\mathbf{Q}$ are three-dimensional orthogonal unit basis vectors along the z^1- and z^2-axis. In perifocal coordinates, the radial coordinate of the orbit is given by

$$r = \frac{a(1-e^2)}{1+e\cos v} \tag{8.16}$$

For the initial conditions, take $v=0$, so the Cartesian position and velocity components are given by

$$\mathbf{r}(0) = (z^1(0), z^2(0), z^3(0)) = (a(1-e), 0, 0) \tag{8.17}$$

$$\mathbf{v}(0) = \left(\frac{dz^1}{dz^0}, \frac{dz^2}{dz^0}, \frac{dz^3}{dz^0}\right)_{s=z^0=0}$$

$$= \left(0, \left(\frac{GM}{ac^2}\frac{1+e}{1-e}\right)^{1/2}, 0\right) \equiv \frac{dz^\beta(0)}{dz^0} \tag{8.18}$$

Note that in Eq. (8.18) I have taken the origin of coordinate time $z^0 = s = 0$, when the satellite is at the periapsis. This is an arbitrary choice made for convenience, and does not affect the results computed over one orbital period.

The transformation between the perifocal coordinates, z^i, and the geocentric equatorial coordinates, x^α, is given by

$$x^0 = z^0 \tag{8.19}$$

$$x^\alpha = d^\alpha_\beta z^\beta, \quad \alpha, \beta = 1, 2, 3 \tag{8.20}$$

where the orthogonal transformation matrix is given by

$$d^\alpha_\beta = \begin{pmatrix} \cos(\Omega)\cos(\omega) - \sin\Omega\sin\omega\cos i & -\cos\Omega\sin\omega - \sin\Omega\cos\omega\cos i & \sin\Omega\sin i \\ \sin\Omega\cos\omega + \cos\Omega\sin\omega\cos i & -\sin\Omega\sin\omega + \cos\Omega\cos\omega\cos i & -\cos\Omega\sin i \\ \sin\omega\sin i & \cos\omega\sin i & \cos i \end{pmatrix}$$

$$(8.21)$$

The 4-dimensional initial conditions are taken as

$$x^0(0) = 0 \tag{8.22}$$

$$x^\alpha(0) = d^\alpha_\beta z^\beta(0) \tag{8.23}$$

$$v^0(0) = 1 \tag{8.24}$$

$$v^\alpha(0) = d^\alpha_\beta \frac{dz^\beta(0)}{dz^0} \tag{8.25}$$

Using the metric in Eq. (8.10), and the normalization of the 4-velocity

$$-1 = g_{ij}v^i v^j \tag{8.26}$$

I find that

$$\frac{dx^0}{ds} = \left[\alpha - \beta\delta_{\kappa\gamma}v^\kappa v^\gamma\right]^{-1/2} \tag{8.27}$$

where

$$\alpha = 1 + \frac{2}{c^2}(V - \phi_o) \equiv 1 + \tilde{\delta} \tag{8.28}$$

$$\beta = 1 - \frac{2}{c^2}V \equiv 1 - \delta \tag{8.29}$$

and both α and β are functions of s through the coordinates $x^\alpha(s)$.

The 4-velocity can then be expressed as

$$v^i(s) = \left[\alpha - \beta\delta_{\kappa\gamma}v^\kappa v^\gamma\right]^{-1/2}(1, v^\alpha) \tag{8.30}$$

So the initial condition on the 4-velocity become

$$v^i(0) = \{\left[\alpha - \beta\delta_{\kappa\gamma}v^\kappa v^\gamma\right]^{-1/2}\}_{s=0}(1, v^\alpha(0)) \tag{8.31}$$

where $v^\alpha(0)$ is given by Eq. (8.25). In all formulas, Greek indices take values $\alpha = 1, 2, 3$, and Latin indices $i = 0, 1, 2, 3$.

Using Eq. (8.13) and (8.12), the rate of change of satellite proper time with coordinate time can be written in terms of the 4-velocity vector components as

$$\frac{ds}{dx^0} = \left[\frac{\alpha}{1 + \beta \delta_{\kappa\gamma} v^\kappa v^\gamma}\right]^{1/2}, \quad \kappa, \gamma = 1, 2, 3 \tag{8.32}$$

Equations (8.11) can be integrated using the initial conditions given in Eq. (8.22)–(8.25). The satellite orbital elements enter the differential equations through Eq. (8.17)–(8.18) and Eq. (8.21).

The goal is to compute the difference between satellite proper time and coordinate time. We also compute the frequency shift (gravitational plus Doppler shift) as observed on Earth, from a satellite transmitting at a known frequency. The absolute orientation of the satellite orbit is not of interest here. Therefore, we set the longitude of the ascending node $\Omega = \pi/2$ and the argument of periapsis $\omega = 3\pi/2$. Setting these two orbital parameters in this way puts the apogee on the negative $x-$axis, on the $z > 0$ half-space. The perigee is on the negative $x-$axis, at $z < 0$. See Figure 8.1.

8.3. Proper Time Minus Coordinate Time for Various Orbits

The difference between proper time τ kept by the satellite clock, and coordinate time $\Delta t = \Delta x^0/c$, kept by clocks on the geoid (in metric given by Eq. (8.10)) is given by

$$\Delta \tau(s) - \Delta t(s) = \frac{1}{c} \int_{s_1}^{s} (1 - \frac{dx^0}{ds}) ds \tag{8.33}$$

where dx^0/ds is given by Eq. (8.32). Equation (8.33) gives the time difference, $\Delta \tau(s) - \Delta t(s)$, at a given value of proper time measured from some epoch, or starting time. We choose the arbitrary starting time $s_1 = c\tau = ct = x^0 = 0$ to be when the satellite is at periapsis.

The gravitational field of the Earth is such that the satellite orbit does not close on itself. In the integration of the differential equations for this model, we have arbitrarily chosen $s = s_1 = 0$ as the point at which a satellite is at periapsis. We take the orbital period to be the value of $s = s_p$ at which the satellite is at its closest approach, as defined by the minimum of the Euclidean distance squared

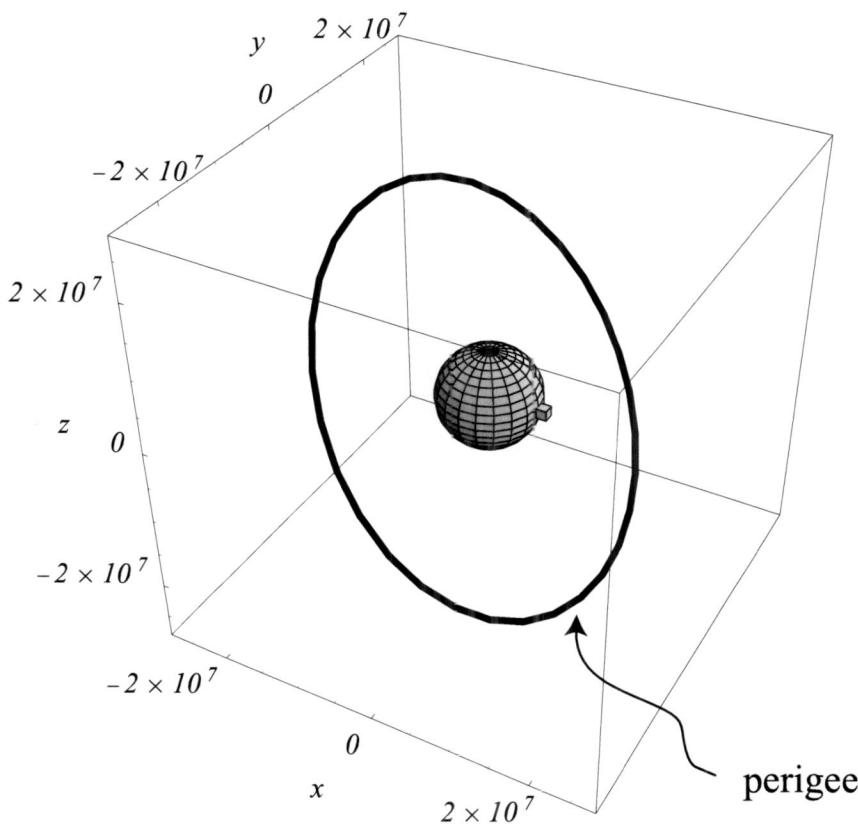

Figure 8.1. The orbit orientation is shown in a real-space 3-dimensional view at time $x^0 = t = 0$, with perigee placed at $z < 0$ and at a maximum value of the x-axis of the orbit. (The perigee is in the lower right side of the figure.) At time $x^0 = t = 0$, the receiver (shown as cube) is on the equator on the surface of the Earth at $(x, y, z) = (R, 0, 0)$, where R is Earth's equatorial radius.

$d^2(s)$:

$$d^2(s) = \sum_{\alpha=1}^{3} (x^\alpha(s) - x^\alpha(s_1))^2 \tag{8.34}$$

For satellites orbiting Earth, we have the small the quantities $v \ll 1$, $\delta \ll$

Table 8.1. Numerical value of the parameters used in the computer program.

c (m/s)	GM (m^3/sec^2)	J_2	R (m)	Earth Rotation Rate $\Omega \frac{radian}{s}$
2.99792458×10^8	3.986005×10^{14}	1.0826800×10^{-3}	6.378137×10^6	$7.2921151467 \times 10^{-5}$

1 and $\tilde{\delta} \ll 1$, and the integrand in Eq. (8.33) can be approximated by

$$1 - \frac{dx^0}{ds} = \frac{1}{2}\left(\tilde{\delta} - v^2\right) - \frac{1}{8}\left[3\tilde{\delta}^2 - v^4 - 4v^2\delta - 2v^2\tilde{\delta}\right] + O(5) \qquad (8.35)$$

where we have dropped terms that we call $O(5)$. Terms of $O(1) \sim 10^{-5}$, and so $v \sim 10^{-5} \sim O(1)$, $\delta \sim 10^{-10} \sim O(2)$, $\tilde{\delta} \sim 10^{-10} \sim O(2)$, for typical Earth orbiting satellites. Terms in the integrand in Eq. (8.33) have been dropped that are of order 10^{-25}. For a satellite with an orbital period that is as large as one day (86400 sec), the error in dropping these terms is on the order of 10^{-20} sec. So our results, within this model, are accurate to 10^{-20} sec per day.

8.4.   Proper Time Minus Coordinate Time: Numerical Results for Various Orbits

A computer program was written in the Mathematica programming language to solve the general relativistic geodesic equations of motion given in Eq. (8.11), subject to the initial conditions on the 4-velocity in Eq. (8.31). The purpose of this program is to illustrate the relativistic effects on satellites in various orbital regimes, and not to predict accurate orbits that include all satellite perturbations. Consequently, the effects of atmospheric drag and solar pressure have been neglected, but in the future these effects could be included. The computer program takes as input the satellite orbit semimajor axis a, eccentricity e, and inclination i, as well as the gravitational constant times the Earth mass, GM, and the value of the earth's quadrupole moment, J_2. The values of these parameters are shown in Table 8.1.

As mentioned above, the orbit is arbitrarily taken to have the longitude of the ascending node $\Omega = \pi/2$ and the argument of periapsis $\omega = 3\pi/2$. Choosing these parameters in this way puts the apogee on the negative x−axis, on the $z > 0$ half-plane, and the perigee on the negative x−axis, at $z < 0$. The computer

Table 8.2. For satellites with given orbital parameters, semimajor axis a, eccentricity e, inclination i, and value of Earth's quadrupole moment J_2, the computed values of proper time minus coordinate time, $\Delta\tau - \Delta t$, are shown per orbital period, and per day. The value per day is the average of this difference over one Earth solar day.

satellite	semimajor axis (meters)	eccentricity	inclination (degrees)	J_2	Δt period (minutes)	$\Delta\tau - \Delta t$ μs per period	$\Delta\tau - \Delta t$ μs per day
LEO[1]	7.3635×10^6	0.00292	82.9	1.0826800×10^{-3}	105.12	-1.290509	-17.678433
LEO[2]	7.3635×10^6	0.00292	82.9	0.0	104.81	-1.301039	-17.875853
GEO[3]	4.2164174×10^7	0.0	0.0	1.0826800×10^{-3}	1435.96	46.4512489	46.5818860
GEO[4]	4.2164174×10^7	0.0	0.0	0.0	1436.0	46.4230537	46.5501514
HEO[5]	2.70365×10^7	0.747194	62.8	1.0826800×10^{-3}	743.08	20.1582623	39.0644760
HEO[6]	2.70365×10^7	0.747194	62.8	0.0	737.37	19.9308525	38.9226991
GPS[7]	2.66965×10^7	0.0017418	55.03	1.0826800×10^{-3}	723.573310	19.438916	38.6858366
GPS[8]	2.66965×10^7	0.0017418	55.03	0.0	723.504421	19.420036	38.6519441

program solves the geodesic equations of motion numerically for the (equatorial geocentric) coordinates $x^i(s)$, $i = 0, 1, 2, 3$, as a function of the proper time s along the orbit. The equations of motion are integrated from $s = 0$ at perigee, to $s = s_p$ at the closest approach to the initial position (approximately one orbital period) which defines the orbital period. At perigee, both the coordinate time and proper time are arbitrarily taken to be equal, and are set to zero. As the satellite progresses in its orbit beyond perigee, proper time deviates from coordinate time, due to relativistic time dilation and gravitational potential effects.

In order to illustrate the numerical values of the effects discussed above, we use a representative satellite from each of four orbital regimes: low-Earth-orbit (LEO), geostationary orbit (GEO), Global Positioning System (GPS) satellite orbits, and highly-elliptical orbit (HEO). The numerical values used in the computer program are shown in Table 8.2. These values were obtained from CelesTrak WWW on the World Wide Web from web site: http://celestrak.com. We compute the difference between elapsed proper time and coordinate time, $\Delta\tau - \Delta t$, where $\Delta\tau$ is the elapsed proper time, and Δt is the elapsed coordinate time, at which the satellite is at perigee. We arbitrarily choose the coordinate time and proper time to be zero when the satellite is at perigee. Then, up to an additive constant, the coordinate time, $t = x^0/c$, where x^0 appears in the metric in Eq. (8.10), is equal to the proper time kept by a reference clock on the geoid, such as the Master Clock at the U.S. Naval Observatory in Washington, D.C.

For each satellite, Table 8.2 shows the computed values of the proper time minus coordinate time difference, $\Delta\tau - \Delta t$, from perigee to (approximate) perigee, accumulated over one orbital period, as determined by solving the geodesic equations of motion, and evaluating Eq. (8.33). The periods of the satellites vary widely, so to compare the magnitude of the effects, the last column in Table 8.2 shows the difference $\Delta\tau - \Delta t$ integrated over one solar day. This value is simply taken to be the average of $\Delta\tau - \Delta t$ over one satellite period, multiplied by the ratio (24 hour)/(satellite period). The calculations in the Table 8.2 are done for a finite value of Earth quadrupole J_2, and for $J_2 = 0$. Comparing the time difference $\Delta\tau - \Delta t$ for $J_2 = 0$ and for finite J_2 gives an estimate of the size of the effect of the Earth's quadrupole on proper time kept by a satellite clock. The coupling to the quadrupole terms is larger for lower altitudes (since the terms proportional to J_2 vary as $1/r^3$) and depends on the inclination of the orbit.

For the low-Earth-Orbit (LEO) satellite, the apogee is at altitude of 1007 km and perigee at 964 km. The difference between proper time and coordinate time, $\Delta\tau - \Delta t = -17.678433$ μs per day, which means that less proper time τ (aboard the satellite) elapses over one solar day than coordinate time t on the geoid . This is due to the predominant effect of time dilation resulting from the high orbital speed of a LEO satellite. Also shown in Table 8.2 is the effect of the non-zero value of the Earth's quadrupole moment, J_2. For the LEO satellite, the finite value of J_2 has the effect of changing the difference $\Delta\tau - \Delta t$ by 197.42 ns.

For high altitude GEO satellite, the difference between proper time and coordinate time, $\Delta\tau - \Delta t$, is of opposite sign to that of a LEO satellite. For a GEO satellite, the proper time elapses faster than coordinate time by 46.5818860 μs per day. The gravitational shift (causing proper time to run fast) dominates the time dilation (which causes proper time to run slow), due to the satellites high altitude and relative slower speed. Also, the GEO satellite is less influenced by the finite size of Earth's quadrupole moment J_2, which is only 31.7346 ns per day, as compared to 197.42 ns for the LEO satellite. Furthermore, for a GEO satellite, the contribution to $\Delta\tau - \Delta t$ from the J_2 quadrupole effect is opposite to that for a LEO satellite.

For the (high-altitude) GPS satellite, apogee is at 20365 km altitude and perigee is at 20272 km altitude. The predominant effect is due to change of gravitational potential, which leads to more elapsed proper time than coordinate time, by 38.6858366 μs per day. Since the altitude of GPS satellites is lower than GEO satellites, the increase of proper time is smaller. For GPS, the effect of nonzero value of J_2 is also similar to the GEO satellite, but has the larger

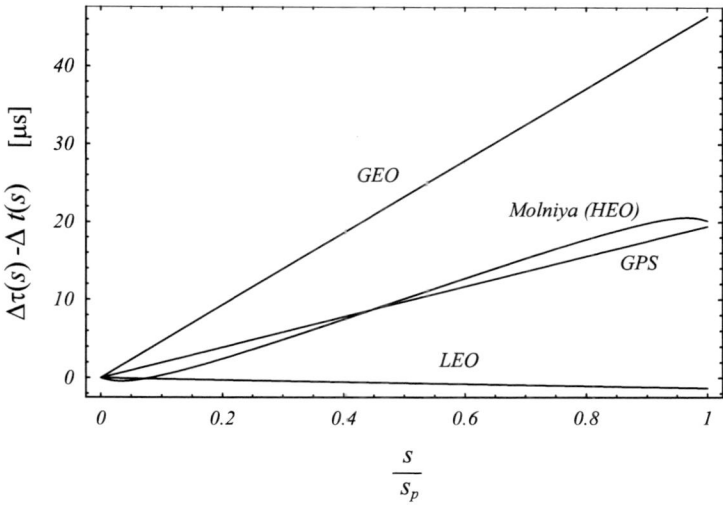

Figure 8.2. The proper time minus the coordinate time, $\Delta\tau(s) - \Delta t(s)$ is shown for the LEO, GEO, HEO (Molniya) and GPS satellites vs. $s/s_p = \tau/\tau_p$, the fraction of proper time period along the orbit. On the abscissa, to the accuracy of the plot, proper time difference is approximately equal to coordinate time difference, $\Delta\tau \sim \Delta t$, where t is coordinate time measured in seconds.

value 33.8925 ns, due to the lower altitude which leads to a greater influence of J_2

The LEO, GEO, and GPS satellites have nearly circular orbits. In contrast, the HEO satellite has an eccentricity of 0.747194, and consequently a perigee of 6835 km (457 km altitude) and apogee 47238 km (40,860 km altitude). This places the HEO satellite both lower and higher than the nearly circular orbit LEO satellite during different portions of its orbit. The altitude of the HEO satellite at perigee and apogee is also lower and higher than the GPS satellite altitude. Consequently, the HEO satellite has proper time running more slowly than coordinate time when it is at low altitude (like the LEO satellite) and it has proper time running faster than coordinate time when its is at high altitude (like GEO and GPS satellites). The net effect for the HEO satellite is that 39.0644760 μs per day more proper time has elapsed than coordinate time. This size of this net effect is similar to the GPS satellite because the HEO satellite spends little time at the low altitude portion of its orbit. The effect of J_2 for the HEO satellite is 141.777 ns, which is 4 times the same effect for the GEO satellite, and 0.7

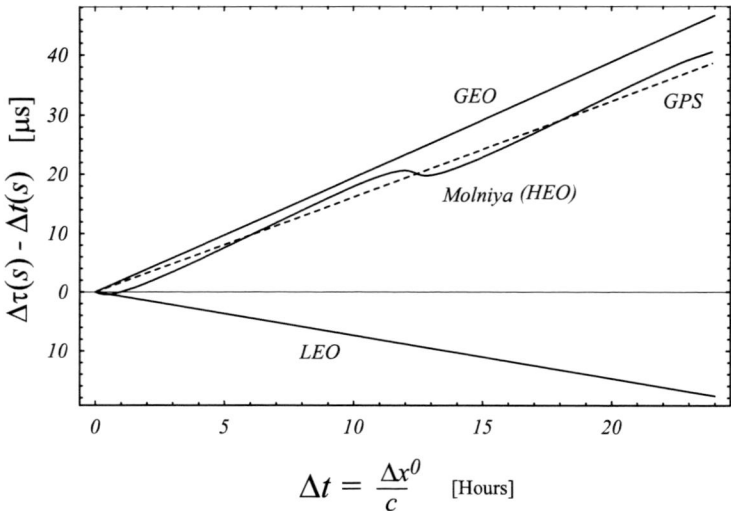

$$\Delta t = \frac{\Delta x^0}{c} \quad \text{[Hours]}$$

Figure 8.3. The proper time minus the coordinate time difference, $\Delta \tau(s) - \Delta t(s)$, is shown for the LEO, GEO, HEO (Molniya) and GPS satellites vs. coordinate time Δt in units of hours. Coordinate time Δt is equivalent to time kept by an ideal reference clock on the Earth's geoid.

times this effect for the LEO satellite.

The results described above are the integrated value of proper time minus coordinate time, over one complete orbital period. It is useful to plot the value of proper time minus coordinate time as a function of the fraction of the orbital period covered, s/s_p, see Figure 8.2. This figure shows that for each satellite, the proper time $\Delta \tau = s/c$ diverges from the coordinate time $\Delta t = \Delta x^0/c$, at a different rate, along its orbit. Note that time is scaled to the orbital period along each orbit, so that $s/s_p = 1$ occurs at the respective period of each orbit. The plot in Figure 8.2 shows a geometrical comparison, however, it does not permit a comparison of the effect in time.

For some practical purposes, it is preferable to look at the proper time minus coordinate time difference, $\Delta \tau - \Delta t$ as a function of coordinate time $\Delta t = \Delta x^0/c$. In this way, the effects are shown as they would appear in real time, say with respect to the USNO Master Clock. Figure 8.3 shows $\Delta \tau - \Delta t$ vs. Δt for the LEO, GEO, HEO (Molniya) and GPS satellites. For each satellite, on the scale of this graph, the plots are almost linear functions. The obvious exception is the HEO

(Molniya) orbit satellite, where, due to its high eccentricity, the time difference is highly nonlinear. Actually, even for the LEO, GEO, and GPS satellites, which have nearly circular orbits, the rate of change of $\Delta\tau(s) - \Delta t(s)$ with time t is non-constant, due to the interaction of the orbital eccentricity, inclination, finite value of J_2, and general relativistic precession effects. These effects are not visible on this graph, but can be significant, depending on the desired level of time synchronization that is required (see Table 1.1) and the time interval over which the synchronization must be maintained (stability). The nonlinear variation of proper time with satellite time can best be seen on a plot of the frequency shift, see the next Subsection.

Since the rate at which proper time diverges from coordinate time is nearly constant for the LEO, GEO, and GPS satellites, these values could be used to compensate for the drift of proper time away from coordinate time, by applying a "factory offset" to the satellite clock frequency, as has been done in the GPS. This could even be done after the orbit is established and the compensation could even be changed periodically. However, inspection of the time synchronization that is required for various applications, see for example Table 1.1, shows that for many applications this compensation is not feasible, due to the non-constant rates (for reasons mentioned above), as well as due to other forces (such as solar pressure and atmospheric drag) that perturb the satellite orbit and lead to proper time changes that cannot be modelled sufficiently accurately. Instead, it is likely that for the most accurate applications of time synchronization, the synchronization will have to be redone periodically, according to the required stability (accuracy required over given time interval).

8.5. Observed Doppler and Gravitational Frequency Shift

When a satellite clock is manufactured, it has a certain prescribed hardware clock rate compared to coordinate time, in some system of coordinates. Alternatively, we can say that the clock, or oscillator in the clock, is calibrated to a given frequency ω_s when it is compared to a local frequency standard. The word local here means that during the calibration procedure, the satellite clock is siting at rest with respect to the frequency standard and that these two devices are at the same gravitational potential (they are co-located and at rest relative to each other). We then refer to the frequency ω_s as the proper frequency of the satellite oscillator or clock.

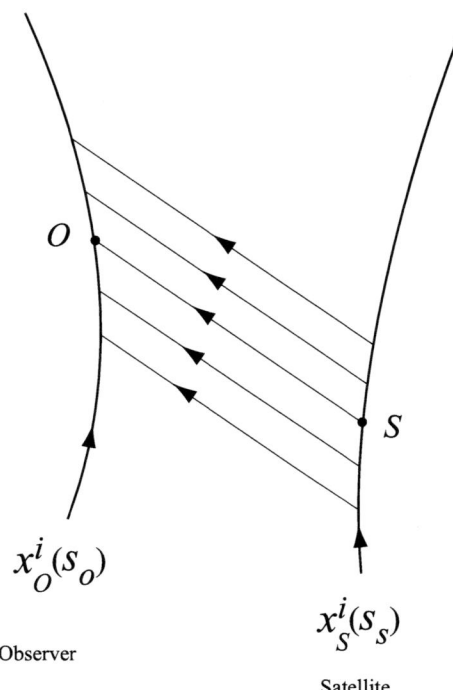

Figure 8.4. The world line of the satellite (transmitter) and observer (receiver) are shown schematically. During satellite proper time ds_s, the satellite clock sends N uniformly-spaced pulses, which are received during observer proper time ds_o. A typical reception event O is connected to an emission event S by a null geodesic.

When the satellite is launched, its clock is assumed to operate normally. In fact, if the frequency standard were placed aboard the satellite, and a comparison was made between the satellite clock and the frequency standard, we would expect to obtain ω_s for the satellite oscillator frequency, to within the calibration accuracy. However, if the frequency standard is located on the Earth's geoid, and the satellite clock sends out electromagnetic pulses, then the frequency of the pulses observed on the geoid (with respect to the frequency standard on Earth) will be ω_o, which is different than the proper frequency of the satellite clock, ω_s. This frequency shift is due to the relative motion of satellite and receiver

(Doppler effect) and also due to the difference in gravitational potential between satellite and receiver. In what follows, we compute this effect.

Consider a satellite orbiting Earth on world line $x_s^i(s_s)$, $i = 0, 1, 2, 3$, where the functions $x_s^i()$ specify the world line of satellite s, and s_s is the parameter related to proper time along the world line, $\tau = s_s/c$. See Figure 8.4. Now consider an observer on a different world line, $x_o^i(s_o)$. The satellite broadcasts a periodic electromagnetic signal, and within the geometric optics approximation, we say that these signals travel along null geodesics, so the reception event O and transmission event S are connected by a null geodesic. In terms of the world function we have

$$\Omega(S, O) = 0 \tag{8.36}$$

where the coordinates of the events are given by

$$S = (x_s^0, x_s^\alpha) \tag{8.37}$$
$$O = (x_o^0, x_o^\alpha), \quad \alpha = 1, 2, 3 \tag{8.38}$$

We compute the frequency measured by the observer, in terms of the frequency transmitted by the satellite. Prior to the measurement, the satellite and observer are assumed to have calibrated their equipment, when they were on a common world line (before launch), as discussed above. We define

$$\omega_o = \text{frequency measured by observer at reception event } O \tag{8.39}$$
$$\omega_s = \text{satellite proper frequency transmitted at event } S, \tag{8.40}$$
$$\text{as measured with respect to frequency standard on board satellite} \tag{8.41}$$

During a proper time $d\tau = ds_o/c$ measured by the observer, there are N cycles of the radiation received. These cycles were sent by the satellite during a proper time ds_s/c, as measured by the satellite clock. Assuming that the signal does not pile up anywhere, e.g., a static space-time, we can say that

$$N = \omega_o \frac{ds_o}{c} = \omega_s \frac{ds_s}{c} \tag{8.42}$$

Note that ds_o is measured at event O and that ds_s is measured at event S and that these two events are arbitrarily separated (but connect by a null geodesic). We can then relate the world lines of the observer and satellite by

$$ds_o^2 = -g_{ij}(O)\,dx_o^i dx_o^j \tag{8.43}$$
$$ds_s^2 = -g_{ij}(S)\,dx_s^i dx_s^j \tag{8.44}$$

where dx_o^i and dx_s^i are increments along the observer and satellite world lines, and $g_{ij}(O)$ and $g_{ij}(S)$ is the metric evaluated at events O and S, respectively. The ratio of frequencies can then be written as

$$\frac{\omega_o}{\omega_s} = \frac{ds_s}{ds_o} = \left[\frac{g_{ij}(S) \frac{dx_s^i}{d\lambda} \frac{dx_s^j}{d\lambda}}{g_{ij}(O) \frac{dx_o^i}{d\lambda} \frac{dx_o^j}{d\lambda}} \right]^{1/2} \tag{8.45}$$

where λ is a common parameter for the two world lines. Note that the frequency ratio in Eq. (8.45) depends on two space-time events, S and O, so this ratio is a two-point tensor, see Section V and Section X-C.

We compute the observer's world line by taking the observer to be stationary on the surface of the rotating Earth. Using ECEF (Earth-centered Earth fixed) coordinates, y^i, $i = 0, 1, 2, 3$, the spatial coordinates of the observer, y^α, are constant. The transformation from ECI Cartesian-like space-time coordinates x^i, to ECEF Cartesian space-time coordinates y^i, is of the form given in Eq. (8.3), with $\omega \to \Omega$, where Ω is the angular frequency of rotation of the Earth, with respect to an ECI coordinate system with a common z-axis. In the ECI coordinates x^i, taking the observer to be located on the y^1-axis on the equator, the observer's world line is then given by choosing $y^1 = R$, $y^2 = 0$, $y^3 = 0$, where R is Earth's equatorial radius.

As a metric in the vicinity of the Earth, we choose Eq. (8.1), with $\bar{x}^0 \to x^0$ and we neglect the Earth's quadrupole moment, taking $J_2 = 0$, so that $V(r, \theta) \to \phi = -GM/r$. The world function for this metric is given by:

$$\Omega(x_1^i, x_2^i) = \frac{1}{2} \eta_{ij} \Delta x^i \Delta x^j + \frac{1}{2} \delta_{ij} \Delta x^i \Delta x^j \frac{2GM}{c^2} \frac{1}{|\mathbf{x}_2 - \mathbf{x}_1|} \log \left(\frac{\tan(\frac{\theta_1}{2})}{\tan(\frac{\theta_2}{2})} \right) \tag{8.46}$$

where $c\Delta t \equiv x_2^0 - x_1^0$, and θ_1 and θ_2 are defined by

$$\cos \theta_a = \frac{\mathbf{x}_a \cdot (\mathbf{x}_2 - \mathbf{x}_1)}{|\mathbf{x}_a||\mathbf{x}_2 - \mathbf{x}_1|}, \quad a = 1, 2 \tag{8.47}$$

We choose the parameter $\lambda = x^0$ in Eq. (8.45), so the ratio of frequencies becomes

$$\frac{\omega_o}{\omega_s} = \left[\frac{1 + \delta_s - (1 - \delta_s) \delta_{\alpha\beta} v_s^\alpha v_s^\beta}{1 + \delta_o - (1 - \delta_o) \delta_{\alpha\beta} v_o^\alpha v_o^\beta} \right]^{1/2} \tag{8.48}$$

where

$$\delta_o = \left[\frac{2}{c^2}\phi\right]_o = \left[-\frac{2GM}{c^2 r}\right]_o \tag{8.49}$$

$$\delta_s = \left[\frac{2}{c^2}\phi\right]_s = \left[-\frac{2GM}{c^2 r}\right]_s \tag{8.50}$$

$$v_o^2 = \delta_{\alpha\beta} v_c^\alpha v_o^\beta \tag{8.51}$$

$$v_s^2 = \delta_{\alpha\beta} v_s^\alpha v_s^\beta \tag{8.52}$$

$$v_o^\alpha = = \frac{dx_2^\alpha}{dx^0} \tag{8.53}$$

$$v_s^\alpha = = \frac{dx_s^\alpha}{dx^0} \tag{8.54}$$

and the subscripts o and s on the square brackets indicate evaluation at events O and S, respectively.

For typical Earth orbiting satellite applications, the δ-terms and velocity terms in Eq. (8.48) are all small: $\delta_o \sim \delta_s \sim v_o^2 \sim v_s^2 \sim O(2)$ where $O(2) \sim 10^{-10}$.

It is then convenient to look at the fractional frequency shift, making the expansion

$$\frac{\omega_o}{\omega_s} - 1 = \frac{\Delta\omega}{\omega_s} = \frac{1}{2}(\delta_s - \delta_o + v_o^2 - v_s^2) - \frac{1}{8}(\delta_s - \delta_o + v_o^2 - v_s^2)^2$$
$$+ \frac{1}{2}\left[(\delta_o - v_o^2)^2 - \delta_o v_o^2 + \delta_s v_s^2 - (\delta_s - v_s^2)(\delta_o - v_o^2)\right] + O(5) \tag{8.55}$$

where $O(5) \sim 10^{-25}$ for typical Earth-orbiting satellites.

As in Eq. (8.11)–(8.32), we use the classical mechanical conditions to integrate the geodesic equations, for the current metric. So the satellite world line is known for the given initial conditions. The satellite world line is parametrized by coordinate time: $x_s^\alpha(x^0)$. The observer's world line, $x_o^\beta(x^0)$, is also parametrized by coordinate time. We know the observer's world line because we assume that the observer is located on the surface of the Earth on the y^1 axis. The calculation is then done as follows. For a given (emission event) satellite coordinate time, x_s^0, and spatial position, x_s^α, we compute the observer's reception event coordinate time x_o^0 by solving:

$$\Omega(x_s^0, x^\alpha(x_s^0), x_o^0, x_o^\beta(x_o^0)) = 0 \tag{8.56}$$

The emission and reception event coordinates, in Eq. (8.37) and (8.38), are then known. These coordinates are used in Eq. (8.55) to plot the observed frequency shift, $\frac{\omega_o}{\omega_s} - 1$ as a function of the reception event time x_o^0.

A computer program in the Mathematica programming language was written to carry out the required computations. The program does not include the (relatively large) atmospheric signal propagation delays, although these effects could be included in the future work. Furthermore, no allowance is made for the Earth obscuring the signal from the satellite to observer.

8.6.   Doppler Plus Gravitational Frequency Shift: Numerical Results for Various Orbits

Proper time on board a satellite is not a linear function of coordinate time. This nonlinear functional dependence can be seen on a plot of the fractional frequency shift, $\Delta\omega/\omega_s$, where $\Delta\omega = \omega_o - \omega_s$, and ω_o is the satellite frequency as observed with respect to a reference oscillator on the geoid, and ω_s is the proper frequency emitted by the satellite–as determined by a frequency standard on board the satellite. See Eq. (8.55) for the definition of $\Delta\omega/\omega$ in terms of orbital parameters. (Here we assume that the satellite frequency standard has been calibrated and that it has not been altered, as has been done in GPS satellites by the "factory offset".) The plots in Figures 8.5–8.9 show the fractional frequency shift, $\Delta\omega/\omega_s$ vs. Δt, where Δt is the elapsed coordinate time measured from time of perigee, for the LEO, GEO, HEO, and GPS satellites. (At perigee we have arbitrarily taken $\Delta t = 0$.)

The frequency shifts, similar to those shown in Figures 8.5–8.9, would be observed at other locations than the Earth's geoid. For example, whenever the proper time is nearly equal to coordinate time, x^0, approximately the same plot would be observed. This is the case for observers in laboratories located at altitude above the geoid. Similar, but more complicated frequency shifts than plotted in Figures 8.5–8.9 would be observed by satellites. Note that the observed frequency shift depends on position and velocity of both observer and satellite. The dependence on spatial position of observer and satellite is due to the fact that space is not homogeneous in a curved space-time, due to the presence of the gravitational field.

The LEO satellite in Figure 8.5 has an overall shift to lower frequency, due predominantly to its high orbital speed. The effect of time dilation dominates the

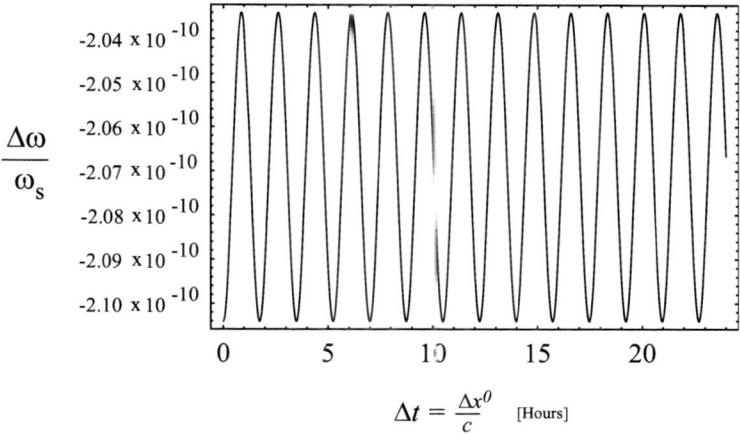

Figure 8.5. The frequency shift (Doppler plus gravitational) as observed from the geoid for the LEO satellite is shown vs. coordinate time $\Delta t = \Delta x^0/c$ in units of hours.

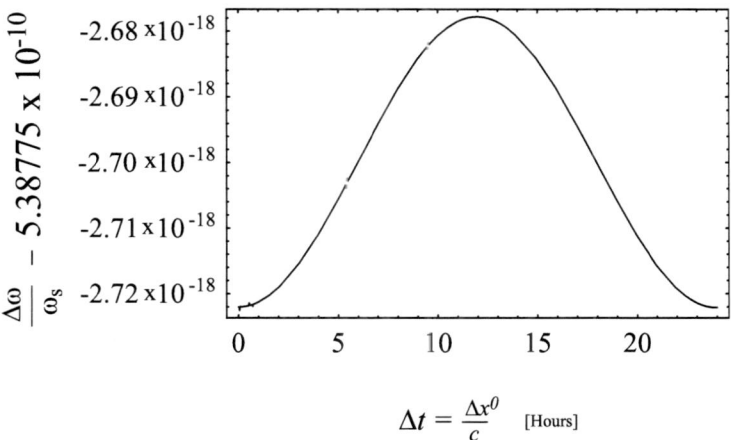

Figure 8.6. The frequency shift (Doppler plus gravitational) as observed from the geoid for the GEO satellite is shown vs. coordinate time $\Delta t = \Delta x^0/c$ in units of hours.

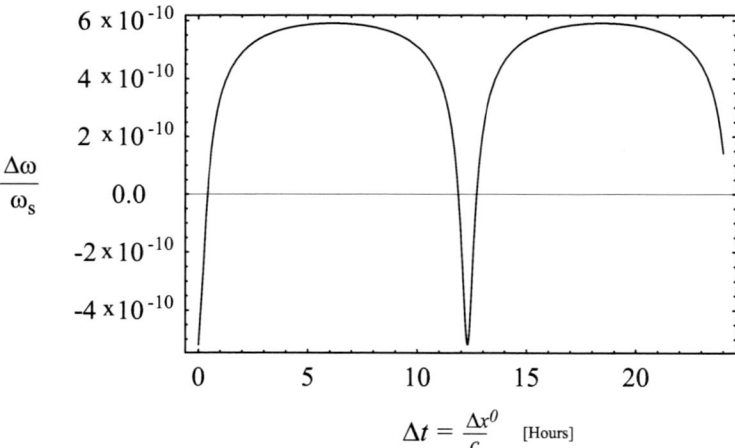

Figure 8.7. The frequency shift (Doppler plus gravitational) as observed from the geoid for the HEO satellite is shown vs. coordinate time $\Delta t = \Delta x^0 / c$ in units of hours.

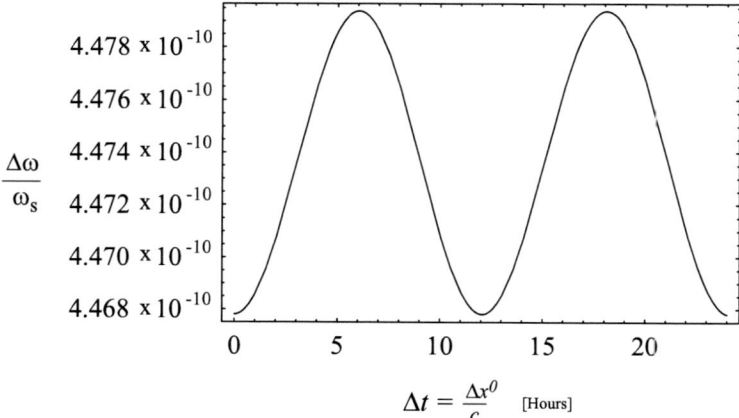

Figure 8.8. The frequency shift (Doppler plus gravitational) as observed from the geoid for the GPS satellite is shown vs. coordinate time $\Delta t = \Delta x^0 / c$ in units of hours.

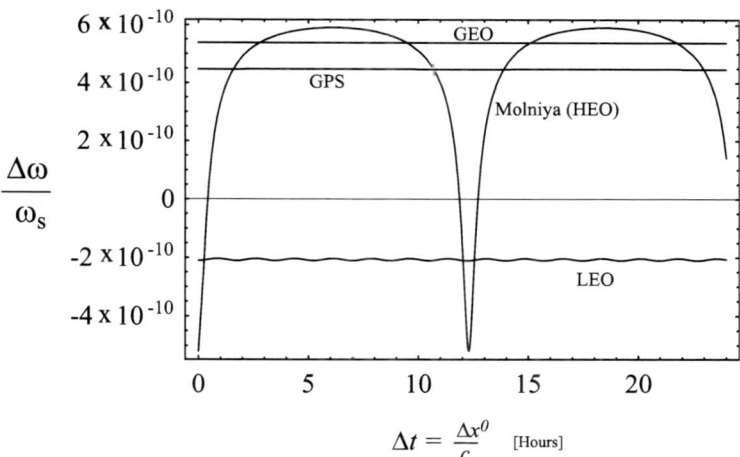

Figure 8.9. The frequency shifts (Doppler plus gravitational) as observed from the geoid is shown vs. coordinate time $\Delta t = \Delta x^0/c$ in units of hours, for the LEO, GEO, HEO (Molniya), and GPS satellites.

effect due to the gravitational potential difference. Superimposed on the overall negative frequency shift, is a small-amplitude rapid variation in the frequency shift due to the satellite's short orbital period This rapid variation is due to the slight orbital eccentricity and the satellite's orbital inclination interacting with the Earth's gravitational field.

The frequency shift of the GEO satellite has an overall positive value of 5.38775×10^{-10}, and is dominated by the gravitational potential frequency shift, causing a shift to higher frequency when observed from the ground, see Figure 8.6. The plot shows the residual frequency shift, after subtraction of 5.38775×10^{-10}, resulting from a slightly non-circular orbit. The satellite parameters have been chosen so that the orbital inclination is zero. The small variation in frequency shift is the result of choosing an initial position and velocity that do not produce a perfectly circular orbit (as may happen when real a satellite is inserted into orbit). In the initial data used here, the result is a small variation in the frequency shift on the order of 2×10^{-18}.

In contrast, the HEO satellite in the Molniya orbit shows a markedly different behavior, see Figure 8.7. The frequency shift is negative for short periods of time when the satellite is near perigee (near $\Delta t = 0$) but is mostly positive

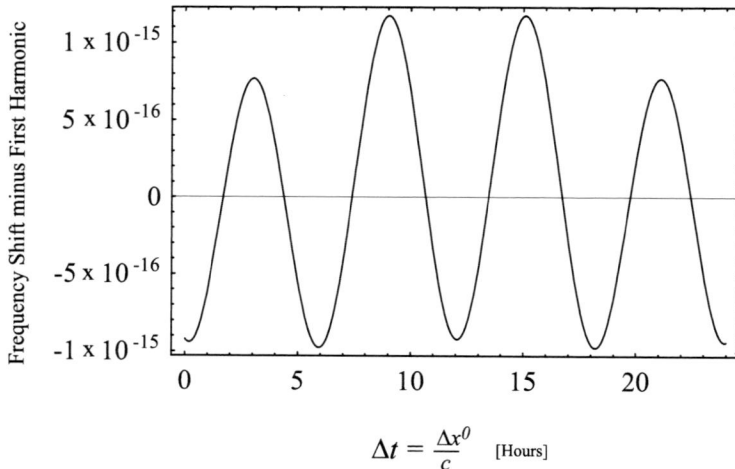

Figure 8.10. A plot of Eq. (8.58) shows that the next higher harmonic, is present in the fractional frequency shift observed on the geoid for the GPS satellite.

for a long time (for approximately 12 hours, roughly between $\Delta t = 0.5$ and $\Delta t = 11.5$ hours) when the satellite spends a long time at high altitude. The HEO satellite makes a rapid transition from the extremes of frequency shift of $\Delta\omega/\omega = +4 \times 10^{-10}$ to -4×10^{-10}, in a time on the order one hour, whereas the satellite period is approximately 12.4 hours.

The frequency shift for the GPS satellite is shown in Figure 8.8. The GPS satellite is in a high, almost circular orbit, with an inclination of 55.3°. The net frequency shift is observed to be positive (higher frequencies are observed on the geoid) with a slight variation due to orbital eccentricity. The net shift is approximately 4.4×10^{-10}, and in the actual GPS system, it is compensated by the factory offset of -4.4×10^{-10}. In addition to this constant frequency shift, the frequency shift also varies at the period of the GPS satellite. Furthermore, plots of frequency shift for all satellites typically have non-zero amplitudes at higher Fourier components than the satellite period. As an example, for the GPS case, we can fit the function

$$A + B\sin(2\pi f t + \phi) \tag{8.57}$$

where A, B, ω' and ϕ are constants, to be fit to the computed function $\Delta\omega/\omega_s$. The values of the best fit parameters are given by $A = 4.4736 \times 10^{-10}$, $B = -5.78749 \times 10^{-13}$, $f = 1.99009$ day^{-1}, and $\phi/2\pi = 0.25011$ day. To demonstrate the presence of higher harmonics, Figure 8.10 shows a plot of the difference

$$\frac{\Delta\omega}{\omega_s} - (A + B\sin(2\pi f t + \phi)) \qquad (8.58)$$

The plot shows that the next harmonic present is at twice the basic GPS period. From Figure 8.10, we see that the amplitude for this next harmonic, which is on the order of 1×10^{-15}, is significantly smaller than for the fundamental frequency. Even though we have chosen a spherical gravity model for the Earth for the frequency shift calculation, higher harmonics are present here because the GPS orbit is slightly eccentric. In general, higher harmonics are present in all satellites because the orbit is eccentric and because the Earth has a non-spherical gravitational field. In the frequency shift calculation presented here, we have neglected the effect of the Earth quadrupole moment, however, it can easily be incorporated into the calculation.

Chapter 9

Geolocation in Curved Space-Time

Location of a source of electromagnetic radiation by using multiple receivers is termed geolocation [3–7]. The source (called the emitter) can radiate a continuous wave (cw) signal or a pulse. In either case, there are two methods commonly employed to locate the emitter: time difference of arrival (TDOA) and frequency difference of arrival (FDOA). The TDOA technique is based on differences of time of flight of the signal from the emitter to each receiver. FDOA is based on the Doppler effect and the difference of frequencies observed by each receiver.

Often the emitter of interest is located on the surface of the Earth. This is a helpful mathematical constraint that reduces the number or receivers needed to locate the emitter. Since satellites are often used to receive signals, and they are costly resources, this constraint helps reduce the number of satellites required for geolocation.

Electromagnetic waves propagate at an (almost) constant velocity in an ECI system of coordinates. However, in this system of coordinates, an emitter that is stationary on the Earth's surface, is rotating with respect to the ECI coordinates, and hence the constraint that the emitter is on the Earth surface is actually a time-dependent constraint (for an Earth model that is not a surface of revolution).

9.1.   Time Difference of Arrival (TDOA) Geolocation

In this section, the world function formalism is used to write the general equations for TDOA geolocation, in an arbitrary system of coordinates, either ECEF or ECI coordinates. Since the world function of space-time is an invariant, the general equations for geolocation are covariant: they are valid in any system of coordinates. I assume that the emitter is located at a space-time event whose coordinates are x_o^i, $i = 0, 1, 2, 3$. Similarly, I assume that three satellites, $s = 1, 2, 3$, receive the emitter signal at space-time events with coordinates x_s^i. The satellites' ephemerides are assumed known, and each satellite carries a clock so the three reception times x_s^0, $s = 1, 2, 3$, are known as well as the spatial coordinates of the reception events. So all satellite space-time coordinates x_s^i are known. The time of emission at the emitter, x_o^0, and the spatial position of the emitter, x_o^α, $\alpha = 1, 2, 3$, constitute four unknowns.

The electromagnetic waves from emitter to each satellite travel on null geodesics that connect the emission event to the three reception events, at each of the three satellites. In an arbitrary system of coordinates, the emitter coordinates are related to the satellite coordinates by the three equations

$$\Omega(x_o^i, x_s^j) = 0 \quad s = 1, 2, 3 \tag{9.1}$$

where Ω is the world function of the space-time. The causality conditions requiring that the reception occur after emission, $x_o^0 < x_s^0$, must also be added for a unique solution.

The constraint that the emitter is on the Earth surface means that the emitter coordinates lie on a 4-dimensional hypersurface, i.e., the two dimensional surface of the Earth sweeps out a 3-dimensional hypersurface (over time) given by:

$$\chi(x_o^i) = 0 \tag{9.2}$$

Equations (9.1) and (9.2) form a system of four nonlinear equations for the four space-time coordinates x_o^i of the emitter. Since this system of equations is nonlinear, it may have multiple solutions, or, no solutions at all.

If the emitter is not on the surface of the Earth, then four satellites (not three) are needed to provide four equations (s=1,2,3,4) of the form of Eqs. (9.1).

As an example of the application of TDOA to locate an emitter, assume the emitter is on the Earth surface and that the space-time surrounding Earth is well-modelled by the Schwarzschild metric. The world function for a Schwarzschild space-time is known [1, 31]. Using Cartesian-like coordinates [1] and solving Eq. (9.1) by iteration leads to

$$x_s^0 = x_o^0 + |\mathbf{x}_s - \mathbf{x}_o| + \frac{GM}{c^2} \left[2\log\left(\frac{\tan(\frac{\theta_o}{2})}{\tan(\frac{\theta_s}{2})} \right) + \cos\theta_o - \cos\theta_s \right] \equiv x_o^0 + \xi(\mathbf{x}_s, \mathbf{x}_o)$$

(9.3)

where $\mathbf{x}_s$ and $\mathbf{x}_o$ are the spatial coordinates of satellites and emitter, and $|\mathbf{x}_s - \mathbf{x}_o|$ is the three-dimensional Euclidean distance between satellite s at reception time and emitter at emission time. When three satellites, $s = 1, 2, 3$, make time of arrival measurements, we can form two independent TDOA equations:

$$x_1^0 - x_2^0 = \xi(\mathbf{x}_1, \mathbf{x}_o) - \xi(\mathbf{x}_2, \mathbf{x}_o)$$

(9.4)

$$x_2^0 - x_3^0 = \xi(\mathbf{x}_2, \mathbf{x}_o) - \xi(\mathbf{x}_3, \mathbf{x}_o)$$

(9.5)

Equations (9.4)–(9.5) are the TDOA equations which contain the effects of the gravitational field on delay of signal propagation (Shapiro time delay, which is on the order of 40 ps). To these two equations, I must add the constraint that the emitter is on the Earth surface. Equations (9.4)–(9.5) are in Schwarschild coordinates, which are essentially ECI coordinates. The transformation from these ECI coordinates, $\mathbf{x} = (x^1, x^2, x^3)$, to rotating ECEF coordinates, $\mathbf{y} = (y^1, y^2, y^3)$, is given by

$$y^0 = x^0$$
$$y^1 = \cos(\frac{\omega}{c}x^0)x^1 + \sin(\frac{\omega}{c}x^0)x^2$$
$$y^2 = -\sin(\frac{\omega}{c}x^0)x^1 + \cos(\frac{\omega}{c}x^0)x^2$$
$$y^3 = x^3$$

(9.6)

In three-dimensional notation, there exists a rotation matrix R, given by Eq. (9.6), that relates the spatial coordinates at a given coordinate time x^0

$$\mathbf{y} = R(x^0) \cdot \mathbf{x}$$

(9.7)

The coordinate time x^0 is taken to be the same in ECI and ECEF coordinates.

For an emission event time x_o^0, the rotating ECEF coordinates are related to the Schwarzschild coordinates by

$$\mathbf{y}_o = R(x_o^0) \cdot \mathbf{x}_o$$

(9.8)

In rotating ECEF coordinates, the Earth's surface is given by

$$f(\mathbf{y}) = 0 \tag{9.9}$$

The constraint in Eq. (9.2) that the emitter is located on the Earth surface is then time-independent and can be written in ECEF rotating coordinates as

$$f(\mathbf{y}_o) = f(R(x_o^0) \cdot \mathbf{x}_o) = 0 \tag{9.10}$$

Note that the unknown time of emission x_o^0 enters explicitly in Eq. (9.10). This time can be eliminated by using Eq. (9.3) (with $s = 1$) in Eq.(9.10), leading to the constraint equation

$$f(R(x_1^0 - \xi(\mathbf{x}_1, \mathbf{x}_o)) \cdot \mathbf{x}_o) = 0 \tag{9.11}$$

The three equations to be solved for the three spatial ECI coordinates $\mathbf{x}_o$ of the emitter are Eq. (9.4)–(9.5) and (9.11). When the coordinates $\mathbf{x}_o$ are found, they are to be substituted into Eq. (9.3) to compute the emission time x_o^0, in terms of the known satellite reception times x_s^0. This time x_o^0 is then used in Eq. (9.8) to compute the emitter ECEF rotating coordinates $\mathbf{y}_o$.

9.2. Doppler Effect in a Gravitational Field

The previous section described the equations needed to carry out geolocation by the technique of TDOA. An alternative and complimentary method is based on measurements of frequency difference of arrival (FDOA).

The frequency that an observer sees emitted by a distant source is not the same as the frequency transmitted by that source. More specifically, the observed frequency differs from the proper (transmitted) frequency because of relative motion between source and observer (Doppler effect) and because of a gravitational potential difference between the position of the source and position of the observer. Most often, the Doppler effect and the gravitational potential effect are treated as separate effects. In a static space-time, there is no difference. However, in a general space-time (not static or stationary), these two effects are inseparable. One expression describes both effects. Below, we follow closely the derivation by Synge [31]. Consider a space-time with an arbitrary gravitational field (not static or stationary). The source of electromagnetic radiation (emitter) travels on world line C' and the observer (satellite with receiver) has

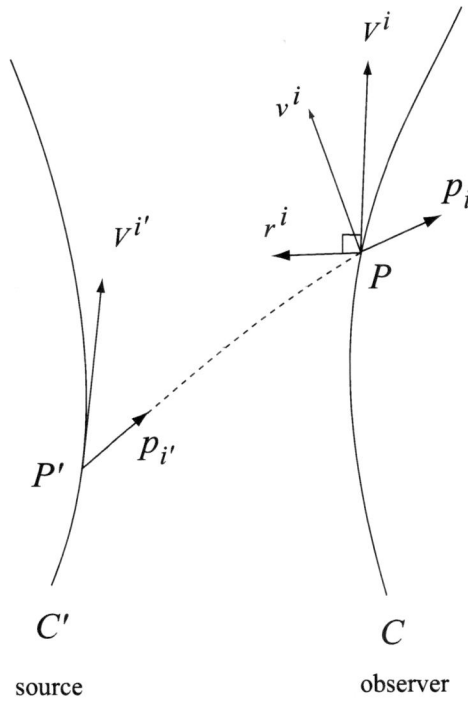

Figure 9.1. The world lines of the emitter C' and receiver C are shown, together with their four velocity, $V^{i'}$ at emission event P', and V^i at reception event P. The photon at emission time has proper momentum $p^{i'}$ and at reception time, has momentum p^i.

world line C, see Figure 9.1. The emitted photon has momentum $p^{i'}$ in the comoving coordinate system of the emitter. In the comoving system of coordinates of the observer (satellite) the photon has momentum p^i. For the comoving system of coordinates of the observer, we define a tetrad $\lambda^i_{(a)}$, $a = 0, 1, 2, 3$. The three spatial components of the tetrad play the role of orthonormal laboratory unit vectors. This tetrad frame does not have to be a Fermi frame (it does not need to be a non-rotating frame). We choose the time-like basis of the tetrad to be the observer 4-velocity $\lambda^i_{(0)} = V^i$. We also use the fact that the three space-like basis vectors of the tetrad are orthogonal to this timelike basis vector $\lambda^i_{(\alpha)} \lambda_{i(0)} = 0$, for $\alpha = 1, 2, 3$, and orthonormal to each other $\lambda^i_{(\alpha)} \lambda_{i(\beta)} = 0$, for $\alpha, \beta = 1, 2, 3$ and $\alpha \neq \beta$. The inner product between the tetrad basis is the

Minkowski matrix:

$$\lambda^i_{(a)} \lambda_{i(b)} = \eta_{(ab)} \tag{9.12}$$

where $\eta_{(ab)} = \text{diag}(-1, +1, +1, +1)$. As usual, tetrad indices are raised and lowered with $\eta_{(ab)}$.

Define the 3-velocity of the source on C', relative to observer on C, by the three invariant projections on the tetrad at P:

$$v_{(\alpha)} = v_i \lambda^i_{(\alpha)} \tag{9.13}$$

However, the velocity $V^{i'}$ of the source and the velocity of the observer V^i cannot be compared directly because they are at different space-time points, P' and P, respectively. Therefore, the 4-velocity vector at P' must be parallel translated to P, defining the vector v_i at P by

$$v_i = g_{ij'} V^{i'} \tag{9.14}$$

where $g_{ij'}$ is the parallel propagator from P' to P. Taking $v^i v_i = -1$, so that v^i is the velocity in units of c, the invariant components on the tetrad basis are related by

$$v^{(0)} = \left(1 + \delta_{\alpha\beta} v^{(\alpha)} v^{(\beta)}\right)^{1/2} = -v_{(0)} \tag{9.15}$$

Then if $v^{(0)} = 1$, the other components $v^{(\alpha)} = 0$, $\alpha = 1, 2, 3$, so that, as compared at point P, the velocity of source and observer are equal.

Now define at point P the unit vector r^i orthogonal to the observer 4-velocity

$$r_i V^i = 0 \tag{9.16}$$

such that the vector r_i lies in the 2-element which contains the tangent vector V^i to C and the tangent vector p_i to the null geodesic $P'P$ that connects the source and observer. Then by Eq. (9.16), $r_i \lambda^i_{(0)} = r_{(0)} = 0$. Define the speed of recession of the source with respect to the observer by

$$v_R = v_i r^i = v_{(\alpha)} r^{(\alpha)} \tag{9.17}$$

since $r^{(0)} = 0$.

Since the curve $P'P$ is a null geodesic, the emitted photon proper momentum $p_{i'}$ is related to the photon momentum p_i at P by parallel transport. Furthermore, the scalar product of two 4-vectors is invariant under parallel transport:

$$p_{i'} V^{i'} = p_i v^i \tag{9.18}$$

The energy of the emitted photon in the comoving frame of reference of the emitter is

$$-E' = p'_i V^{i'} = p_{(0)} v^{(0)} + p_{(\alpha)} v^{(\alpha)} \tag{9.19}$$

The energy of a photon, with momentum p^i, seen in the comoving frame of the observer with 4-velocity V^i, is given by

$$E = -p_i V^i = -p_i \lambda^i_{(0)} = -p_{(0)} = p^{(0)} \tag{9.20}$$

From the definition of unit vector r^i, i.e., that it exists in the 2-element that contains the tangent V^i at P to C, and the null geodesic tangent vector p_i, and the fact that p_i is a null vector, the vector p_i can be written as a decomposition

$$p^i = \alpha(V^i - r^i) \tag{9.21}$$

where $r_i r^i = 1$ and the magnitude α is to be determined. Multiplying Eq. (9.21) by V_i, I find that $p^i V_i = -\alpha = -E$. Therefore, Eq. (9.21) can be written as

$$p^i = E(V^i - r^i) \tag{9.22}$$

From Eq. (9.19), and using $p_{(0)} = -E$,

$$E' = E v^{(0)} - p_{(\alpha)} v^{(\alpha)} \tag{9.23}$$

The term $p_{(\alpha)} v^{(\alpha)} = p_i \lambda^i_{(\alpha)} v^{(\alpha)}$ and using Eq. (9.22) for p_i leads to

$$
\begin{aligned}
p_{(\alpha)} v^{(\alpha)} &= E(V_i - r_i) \lambda^i_{(\alpha)} v^{(\alpha)} &\tag{9.24} \\
&= E\left[V_i \lambda^i_{(\alpha)} - r_i \lambda^i_{(\alpha)} \right] &\tag{9.25} \\
&= -E r_{(\alpha)} v^{(\alpha)} = -E v_R &\tag{9.26}
\end{aligned}
$$

since $V_i \lambda^i_{(\alpha)} = 0$ because the frame is orthogonal, and we used Eq. (9.17) for the definition of the recessional velocity v_R. Now use Eq. (9.26) in Eq. (9.23) for $p_{(\alpha)} v^{(\alpha)}$ and Eq. (9.15) for $v^{(0)}$ to obtain

$$E = \frac{E'}{(1 + v^2)^{1/2} + v_R} \tag{9.27}$$

where

$$v^2 = \delta_{\alpha\beta} v^{(\alpha)} v^{(\beta)} \tag{9.28}$$

and $\alpha, \beta = 1, 2, 3$. In Eq. (9.27), we have the following definitions:

E=photon energy observed by satellite at P

E'=photon energy emitted in (proper) comoving frame of emitter at P'

v^2=square of speed in terms of components on tetrad, defined by Eq. (9.28)

v_R = recessional velocity defined by Eq. (9.17)

Equation (9.27) shows that the frequency of an electromagnetic signal observed by a satellite, E, is related to the frequency emitted, E', by motional (velocity) effects. Of course, the role of the gravitational field in producing a frequency shift (gravitational red shift effect) is contained in Eq. (9.27) in the geometry of the curved space-time. The significance of Eq. (9.27) is that it is a general expression, that is valid in an arbitrary space-time and does not assume a static or stationary space-time. However, Eq. (9.27) does assume that geometrical optics is valid because the geodesic law for light propagation was used.

9.3. Observed Doppler Shifts

The observed energy of a photon, E, is related to the observed angular frequency ω, by $E = \hbar\omega$. In a system of coordinates where the photon 4-momentum is p_i, and the observer 4-velocity is V^i, the observed photon frequency is given by $E = \hbar\omega = -p_i V^i$. Similarly, the energy of the photon emitted by the emitter, in the comoving frame of the emitter, is related to the frequency by $E' = \hbar\omega' = -p_{i'} V^{i'}$, where $p_{i'}$ is the 4-momentum at the emission event and $V^{i'}$ is the 4-velocity of the emitter.

The fractional shift in frequency, between the emitted and received frequencies, due to motional doppler effects and gravitation is given by

$$\frac{E' - E}{E'} = \frac{p_{i'} V^{i'} - p_i V^i}{p_{i'} V^{i'}} \tag{9.29}$$

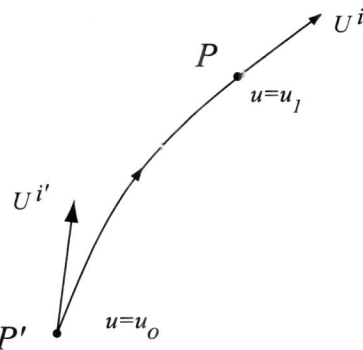

Figure 9.2. A geodesic path is shown between points P' and P, with unit vectors $U^{i'}$ and U^i at the end points.

The right side of Eq. (9.29) can be written in terms of the derivative of world function of the space-time. The world function connects two points, P' and P, by a geodesic. The geodesic connecting P' and P may be expressed parametrically as $x^i(u)$, for $i = 0, 1, 2, 3$, for $u_0 \leq u \leq u_1$, where $P : x^i(u_o)$ and $P : x^i(u_1)$.

The world function is a two-point scalar, depending on points P' and P. The covariant derivatives of the world function with respect to point P' and P are given by

$$\Omega_{i'} = \Omega_{i'}(P', P) = \frac{\partial \Omega}{\partial x^{i'}} = -(u_1 - u_o)U_{i'} \tag{9.30}$$

$$\Omega_i = \Omega_i(P', P) = \frac{\partial \Omega}{\partial x^i} = (u_1 - u_o)U_i \tag{9.31}$$

The photon momentum, $p_{i'}$ at point P', is related to the photon momentum, p_i at point P, by parallel transport. Also, the photon momentum is in the direction of the tangent to the geodesic. Also, the magnitude of a vector (photon momentum) is constant under parallel transport. Therefore, the momentum is proportional to the tangents at point P' and at P

$$U_{i'} = \alpha p_{i'} \tag{9.32}$$

$$U_i = \alpha p_i \tag{9.33}$$

where α is the same proportionality constant. Using Eq.(9.30)–(9.33), the frequency difference in Eq. (9.29) can be written as

$$\frac{\omega' - \omega}{\omega'} = \frac{\Omega_{i'}V^{i'} + \Omega_i V^i}{\Omega_{i'}V^{i'}} \tag{9.34}$$

Eq. (9.34) gives the relation between the frequency ω observed by a satellite at the reception event P and the frequency of the emitted electromagnetic signal at the emission event P'. The satellite has 4-velocity V^i and the emitter has 4-velocity $V^{i'}$. The frequency ω' is the proper frequency of the electromagnetic signal broadcast by the emitter in its comoving frame, i.e., the frame in which the emitter is at rest. Eq. (9.34) contains the effect of the gravitational potential (gravitational red shift) as well as the effect associated with the relative motion of the source and observer, which is usually called the Doppler effect.

The significance of the expression in Eq. (9.34) is that it can be evaluated in any system of coordinates (any reference frame). For example, Eq. (9.34) can be evaluated in inertial or rotating coordinates. The key ingredient to carrying out the calculation is that the world function must be computed for the given space-time. This has already been done for the Schwarzschild space-time, which models the Earth as a sphere [1, 31]. A similar calculation of the world function should be done to include the effects of the Earth's quadrupole potential J_2. However, the gravitational effects due to J_2 are three orders of magnitude smaller than the monopole terms, and for some applications, may be negligible when dealing with computations of signal propagation time or frequency shift. For example, if the time delay due to the monopole contribution is of the order of 40 ps, then the effect of J_2 is expected to be three orders of magnitude smaller. Furthermore, the doppler effect is not a cumulative effect such as time dilation, so these terms do not increase in size with time. Therefore, for many purposes the world function for the Schwarzschild space-time is sufficient.

9.4. Frequency Difference of Arrival (FDOA) Geolocation

In this section, we derive the basic relations for geolocation by frequency difference of arrival (FDOA) in a curved space-time. Geolocation means that measurement of the frequency of an emitter by several satellites (in relative motion, and at known positions in space-time) can be used to locate the emitter in space-time. Here we will take into account the frequency shifts due to the relative motion of satellite and emitter, as well as due to the gravitational potential effect. When satellites in different orbital regimes, e.g., LEO, GEO and HEO, are combined, the gravitational potential differences can lead to significant errors in emitter positions. We neglect the (relatively large) effects that can result from atmospheric propagation delays.

As derived in section X, subsection C, Eq. (9.34) gives the basic expression that connects the frequency emitted (proper frequency) with the frequency observed, taking account the relative motion as well as different gravitational potentials. Assume that the emitter sends a signal at event $P_0 = (x_o^0, x_o^\alpha)$ and that this signal is received by a satellite at event $P_s = (x_s^0, x_s^\alpha)$. The emitted frequency as measured at the emitter using a reference oscillator is ω_0. The frequency observed at the satellite, ω_s, is observed by using a reference oscillator at the satellite identical to that used at the emitter to determine the emitter's frequency. The fractional frequency difference can be written in terms of the world function and the 4-velocities of the emitter and satellite receiver:

$$\frac{\omega_o - \omega_s}{\omega_o} = \frac{\Omega_{i_o} V^{i_o} + \Omega_{i_s} V^{i_s}}{\Omega_{i_o} V^{i_o}} \equiv R \tag{9.35}$$

where for practical applications $R \ll 1$. In Eq. (9.35), we have the following definitions

$$\Omega_{i_o} = \frac{\partial \Omega(P_o, P_s))}{\partial x_o^i}$$

$$\Omega_{i_s} = \frac{\partial \Omega(P_o, P_s))}{\partial x_s^i}$$

$$V^{i_o} = \frac{dx_o^i}{ds} = \text{4-velocity of emitter}$$

$$V^{i_s} = \frac{dx_s^i}{ds} = \text{4-velocity of satellite}$$

$$\omega_o = \text{proper frequency of emitter, with 4-velocity } V^{i_o}$$

$$\omega_s = \text{frequency of emitter, as observed at}$$
$$\text{satellite with 4-velocity } V^{i_s}$$

The frequency observed by a given satellite can be written as

$$\omega_s = \omega_o(1 - R) \tag{9.36}$$

where R depends on the 4-velocities of emitter and receiver, and the world function of space-time. For the Minkowski metric given in Appendix B, explicit evaluation of Eq. (9.35) gives

$$\omega_s = \omega_o = \frac{\gamma_s(1 - \mathbf{n} \cdot \mathbf{v}_s)}{\gamma_o(1 - \mathbf{n} \cdot \mathbf{v}_o)} \tag{9.37}$$

where $\mathbf{v}_o$ and $\mathbf{v}_s$ are the velocities of the emitter and satellite (in units of v/c), and the unit $\mathbf{n}$ vector is defined as $\mathbf{n} = (\mathbf{r}_s - \mathbf{r}_o)/|\mathbf{r}_s - \mathbf{r}_o|$, where $\mathbf{r}_o$ and $\mathbf{r}_s$ are the spatial position of emitter and satellite at points of emission and reception.

The measured quantities are frequency differences, which can be formed from difference of satellite frequency given in Eqs. (9.36). We do not give the final complicated expressions here for frequency differences. However, oscillator frequency errors clearly contribute to errors in estimated emitter position. For example, preliminary analysis shows that the order of magnitude error in emitter location is given by

$$|\delta\mathbf{x}_o| = \frac{\delta\omega}{\omega} \frac{|\mathbf{x}_s - \mathbf{x}_o|}{|\mathbf{v}_s - \mathbf{v}_o|/c} \tag{9.38}$$

where $\mathbf{x}_o$, $\mathbf{v}_o$ and $\mathbf{x}_s$, $\mathbf{v}_s$ are the position and speed of the emitter and receiver, respectively, and $\frac{\delta\omega}{\omega}$ is the fractional frequency error. For example, for the typical values of satellite speed $v_s \sim 10^{-5}$, and satellite oscillator frequency error $\frac{\delta\omega}{\omega} \sim 10^{-10}$, the order of magnitude of emitter position error is $|\delta\mathbf{x}_o| \sim 100$ m. A frequency error on the order of 10^{-10} can also be expected to occur when a LEO and a GEO satellite are used together for geolocation without compensating for the gravitational frequency shift. In such as case, the use of two satellites can lead to a similar position error on the order of 100 m. Detailed investigation of Eq. (9.36) for high-accuracy geolocation is left for future work.

Chapter 10

Summary

In this article we have considered the elements of the general problem of navigation in space-time, as well as the restricted problem of clock synchronization, within the context of a metric theory of gravity, such as general relativity. In most real applications, such as the GPS, a user is interested in determining his space and time coordinates, rather than just time. General relativity deals with the effect of motion and gravitational potential differences on clocks, and it highlights features of the space-time navigation problem that must be included in any (classical or quantum) theory of navigation (or clock synchronization). Several quantum mechanical schemes have been proposed to synchronize clocks [21–30]. At the present time, the effects of motion and gravitational potential differences have not been explicitly incorporated into these quantum approaches to clock synchronization. Clearly the magnitude of the relativistic effects is such that it must be considered in future quantum approaches to navigation. At the present, there is no quantum theory of navigation in space-time (analogous to the classical theory in Ref. [2]) which permits a user to determine their four space-time coordinates.

Acknowledgments

The author is grateful to Pete Hendrickson for numerous helpful discussions. This work was supported by the Advanced Research and Development Activity (ARDA).

Appendix

A. Significance of Accurate Clock Synchronization for Geolocation

Consider the problem of locating an emitter of electromagnetic radiation located at a space-time point $(t, \mathbf{r})$ near the Earth's surface. The time of emission and position of this emitter can be determined from signals received at fours satellites, located at reception events $(T_a, \mathbf{R}_a)$, $a = 1, 2, 3, 4$, see Fig. 10.1.

In an inertial system of coordinates (such as the ECI frame), the time and position of the emitter can be computed by solving the four equations

$$c(T_a - t) = |\mathbf{r} - \mathbf{R}_a|, \quad a = 1, 2, 3, 4 \tag{A.1}$$

for $(t, \mathbf{r})$, where the satellite positions, $\mathbf{R}_a$, and reception times, T_a, are known, and c is the speed of light in vacuum. In this Appendix, for simplicity, we neglect the distinction between proper time and coordinate time.

By taking cyclic differences of Eqns. (A.1), we cancel out the emission event time t and obtain time difference of arrival (TDOA) equations

$$
\begin{aligned}
c(T_1 - T_2) &= |\mathbf{r} - \mathbf{R}_1| - |\mathbf{r} - \mathbf{R}_2| \tag{A.2} \\
c(T_2 - T_3) &= |\mathbf{r} - \mathbf{R}_2| - |\mathbf{r} - \mathbf{R}_3| \tag{A.3} \\
c(T_2 - T_4) &= |\mathbf{r} - \mathbf{R}_3| - |\mathbf{r} - \mathbf{R}_4| \tag{A.4}
\end{aligned}
$$

where the measured quantities are the time differences $T_a - T_{a+1}$ at the satellites. The Eqns. (A.2) can be solved for the coordinates of the emitter $\mathbf{r}$, in an inertial coordinate system. The value of $\mathbf{r}$ must then be substituted into one Eqns. (A.1) to obtain the emission time t. The coordinates of the emitter, $\mathbf{r}$, can then be transformed to geocentric (rotating with the Earth) coordinates, $\mathbf{r}'$, by a time-dependent rotation

$$\mathbf{r}' = \underline{\mathbf{D}}(t) \cdot \mathbf{r} \tag{A.5}$$

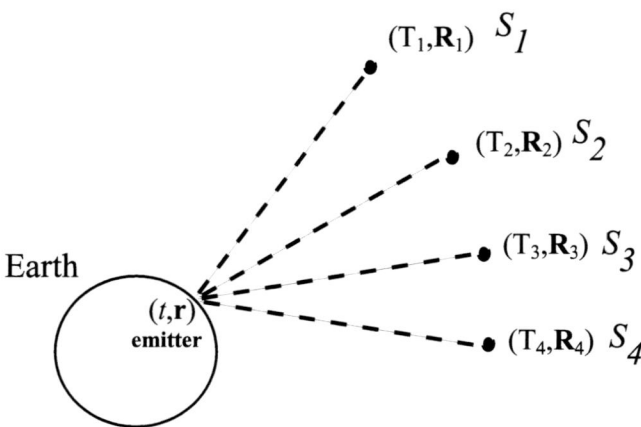

Figure 10.1. The time and position of an electromagnetic emitter is shown at a space-time point $(t, \mathbf{r})$, together with satellites, S_a, located at $(T_a, \mathbf{R}_a)$, $a = 1, 2, 3, 4$.

where $\underline{\mathbf{D}}(t)$ is the transformation matrix.

A similar problem occurs if we know that the emitter is located on the surface of the Earth, so that we have the constraining equation

$$|\mathbf{r}| = r_E \qquad (A.6)$$

where r_E=6378 km is the Earth's radius. In this case, a minimum of three satellites are sufficient to locate the emitter. The first two of Eqns. (A.2) are solved together with the constraining Eqns. (A.6) for $\mathbf{r}$. Then the geocentric (rotating with the Earth) coordinates, $\mathbf{r}'$, are found by the time-dependent rotation in Eqns. (A.5).

The question then arises: How does the accuracy of the synchronization of the clocks affect the accuracy of geolocation? Assume that the clocks of two satellites are synchronized, and that the satellite positions are known. The maximum TDOA occurs for satellites separated by an angle ϕ given by (see Figure 10.2)

$$\sin\phi = l_2/r_s \qquad (A.7)$$

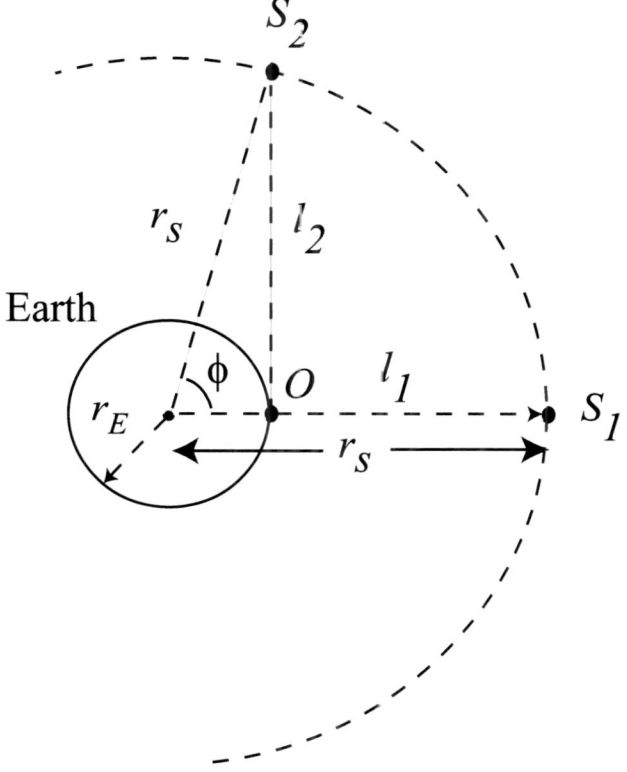

Figure 10.2. The orientation of an electromagnetic emitter O is shown with respect to two satellites S_1 and S_2.

The path length difference between these two satellites is

$$\Delta l = l_2 - l_1 = \sqrt{r_s^2 - r_E^2} - (r_s - r_E) \tag{A.8}$$

The maximum time difference of arrival (TDOA) for a signal is

$$\Delta t = \frac{\Delta l}{c} = \frac{l_2 - l_1}{c} = \frac{\sqrt{r_s^2 - r_E^2} - (r_s - r_E)}{c} \tag{A.9}$$

For geosynchronous satellites, where $r_E =$6378 km and $r_s =$42,164 km, the maximum (TDOA) is $\Delta t =$19.64 ms, and it occurs when one satellite has a longitude that is the same as the emitter and the other satellite is at $\phi =$81.3°, see

Figure 10.2. Due to geometric constraints, the whole range difference is contained in the maximum time delay of 19.64 ms. If the clocks in the two satellites are synchronized only to an accuracy of, say 10 ns, then the whole range difference is compressed into a time delay of 19.64 ms. The fractional range error, and consequently the order of magnitude of the position error (neglecting geometric dilution of precision (GDOP) factors) is given by

$$\Delta x \sim \Delta l = (l_2 - l_1)\frac{10\text{ns}}{19.64\text{ms}} \sim 3000\text{m} = 1.5\text{nm} \tag{A.10}$$

An improvement in the clock synchronization, say by three orders of magnitude, translates directly into three orders of magnitude improvement in position accuracy, resulting in position accuracy of the order of 3 m.

In the discussion in this appendix, we have neglected atmospheric (tropospheric and ionospheric) time-delay effects, accuracy of computational correlation algorithms, and noise in the receiving system. An implicit assumption is also made that the system of coordinates (inertial reference frame) is accurate, and that the time-dependent transformation between inertial and geocentric (rotating) coordinates is accurately known.

B. Conventions and Notation

Where not explicitly stated otherwise, we use the convention that Roman indices, such as found on space-time coordinates x^i, take the values $i = 0,1,2,3$, and Greek indices take values $\alpha = 1,2,3$. Summation is implied over the range of any index when the same index appears in a lower and upper position. In some cases, summation over Greek indices is implied when indices both appear in upper position, such as in $x^\alpha dx^\alpha$.

If x^i and $x^i + dx^i$ are two events along the world line of an ideal clock, then the square of the proper time interval between these events is $d\tau = ds/c$, where the measure ds is given in terms of the space-time metric as $ds^2 = -g_{ij}\,dx^i\,dx^j$. I choose g_{ij} to have the signature $+2$. When g_{ij} is diagonalized at any given space-time point, the elements can take the form of the Minkowski metric given by $\eta_{00} = -1$, $\eta_{\alpha\beta} = \delta_{\alpha\beta}$. In discussion of an observer carrying a tetrad, the Minkowski matrix, $\eta_{(ab)} = \eta_{ab}$, is used.

C. Acknowledgement

This work was funded by the National Reconnaissance Office (NRO). Portions of this work were completed while the author was at the U. S. Army Research Laboratory, 2800 Powder Mill Road, Adelphi, Maryland, USA 20783-1197.

References

[1] T. B. Bahder, "Navigation in curved space-time", *Am. J. Phys.* **69**, 315 (2001).

[2] T. B. Bahder, "Relativity of GPS Measurement", *Phys. Rev. D* **68**, 063005 (2003).

[3] K. C. Ho and Y. T. Chan, "Solution and Performance Analysis of Geolocation by TDOA", *IEEE Trans. Aerospace and Electronic Systems,* **29**, 1311 (1993).

[4] B. T. Fang, "Comments on "Analysis of Geolocation by TDOA" ", *IEEE Trans. Aerospace and Electronic Systems,* **31**, 510 (1995).

[5] G.H. Niezgoda and K.C. Ho, "Goelocation by combined range difference and range rate measurements", In *Proceedings of the 1994 International Conference on Acoustics, Speech*, and Signakl Processing, Adelaide, South Australia, April 1994, vol. II, 357-360.

[6] K. C. Ho and Y. T. Chan, "Geolocation of a Known Altitude Object from TDOA and FDOA Measurements", *IEEE Trans. Aerospace and Electronic Systems*, **33**, 770 (1997).

[7] T. Pattison and S.I. Chou, "Sensitivity Analysis of Dual-Satellite Geolocation", *IEEE Trans. Aerospace and Electronic Systems,* **36**, 56 (2000).

[8] B. F. Burke, "Introduction to Orbiting VLBI", *Adv. Space Res.* **11**, 349 (1991).

[9] Steyskal, H.; Schindler, J.K.; Franchi, P.; Mailloux, R.J., "Pattern synthesis for TechSat21–a distributed spacebased radar system", *Aerospace Conference, 2001, IEEE Proceedings*, Vol. 2, p.725 (2001).

[10] O. J. Sovers, and J. L. Fanselow, "Astrometry and geodesy with radio inter-
      ferometry: experiments, models, results", *Rev. Mod. Phys.* **70**, 1393-1454
      (1998).

[11] B. Boverie, "Optimization of the geometry of distributed aperture re-
      ceivers", Ph. D. Thesis, Univ. of Texas at Austin, (1970).

[12] See for example, K. C. Overman, K. A. Leahy, T.W. Lawrence, R.J.
      Fritsch, "The Future of Surface Surveillance–Revolutionizing the View of
      the Battlefield", *IEEE International Radar Conference*, Alexandria, Vir-
      ginia, USA, 7-12 May 2000.

[13] Hodge, C.C.; Klein, H.A.; Knight, D.J.E.; Maleki, L., "Ultra-stable Op-
      tical Frequencies for Space", *Proceedings of the 1999 Joint Meeting of
      the European Frequency and Time Forum and the IEEE International Fre-
      quency Control Symposium,* Besancon, France; 13-16 April 1999, 663-6
      vol.2.

[14] D. M. Le Vine and J.C. Good, "Aperture Synthesis for Microwave Ra-
      diometers in Space", *NASA Technical Memorandum 85033*, August 1983,
      Goddard Space Flight Center, Greenbelt, Maryland.

[15] D. M. Le Vine, "The sensitivity of Synthetic Aperture Radiometers for
      Remote Sensing Applications from Space", *Radio Science* **25**, 441 (1990).

[16] R.J.Sedwick,T.L. Hacker and D.W. Miller, "Optimum Aperture Placement
      for a Space-Based Radar System Using Separated Spacecraft Interferom-
      etry", *AIAA Guidance, Navigation, and Control Conference and Exhibit*,
      9-11 August 1999, Portland, Oregon.

[17] M.M. Colavita, J.P. McGuire, R.K. Bartman, G. H. Blackwood, R. A.
      Laskin, K. Lau, M. Shao, and J. W. Yu, "Separated Spacecraft Interferom-
      eter concept for the New Millennium Program", *SPIE Conference 2807:
      Space Telescopes and Instruments*, Denver, CO, 6-7 August 1996.

[18] For example, see the web sites: http://www.vsop.isas.ac.jp, and
      http://www.vsop.isas.ac.jp/vsop2/.

[19] D. Bouwmeester, A. Ekert, A. Zeilinger, *The Physics of Quantum Infor-
      mation: Quantum Cryptography, Quantum Teleportation, Quantum Com-
      putation,* Springer, New York (2000).

[20] M. A. Nielson and I. L. Chuang, *Quantum Computation and Information*, Cambridge University Press, New York (2000).

[21] C.K. Hong, Z.Y. Ou, and L. Mandel, "Measurement of subpicosecond time intervals between two photons by interference", *Phys. Rev. Lett.* **59**, 2044-6, (1987).

[22] I. L. Chuang, "Quantum Algorithm for Distributed Clock Synchronization", *Phys. Rev. Lett.* **85**, 2006 (2000), and also in quant-ph/0004105.

[23] R. Jozsa, D. S. Abrams, J. P. Dowling, and C. P. Williams, "Quantum Clock Synchronization Based on Shared Prior Entanglement", *Phys. Rev. Lett.* **85**, 2006 (2000).

[24] U. Yurtsever and J. P. Dowling, "A Lorentz-invariant Look at Quantum Clock Synchronization Protocols Based on Distributed Entanglement", quant-ph/0010097.

[25] E. A. Burt, C. R. Ekstrom, T. B. Swanson, "Comment on "Quantum Clock Synchronization Based on Shared Prior Entanglement"", *Phys. Rev. Lett.* **87**, 129801 (2001); "A Reply to "Quantum Clock Synchronization"", quant-ph/0007030.

[26] R. Jozsa, D. S. Abrams, J. P. Dowling, and C. P. Williams, "Jozsa et al. Reply", *Phys. Rev. Lett.* **87**, 129802 (2001).

[27] J. Preskill, "Quantum Clock Synchronization and Quantum Error Correction", quant-ph/0010098.

[28] V. Giovannetti, S. Lloyd, and L. Maccone, *Nature* **412**, 417-419 (2001);

[29] T. B. Bahder and W. M. Golding, "Clock synchronization based on second-order coherence of entangled photons", submitted for publication.

[30] A. Valencia, G. Scarcelli and Y. Shih, "Ultra-high accuracy nonlocal timing and positioning–beyond the classical limit", submitted for publication.

[31] J. L. Synge, *Relativity: The General Theory*, North-Holland Publishing Co., New York, (1960).

[32] L. D. Landau and E. M. Lifshitz, *Classical Theory of Fields*, Pergamon Press, New York, Fourth Revised English Edition, (1975).

[33] For the conventions associated with the metric g_{ij} used in this report, see Appendix A.

[34] For a discussion of this topic, see for example p.58 in V. A. Brumberg, *Essential Relativistic Celestial Mechanics*, Adam Hilger, New York, (1991).

[35] See, for example, C. M. Will, "Theory and Experiment in Gravitational Physics" (Cambridge University Press, New York, 1993), revised ed., pp. 67-85.

[36] C. M. Will,"The Confrontation between General Relativity and Experiment: A 1998 Update", *Lecture notes from the 1998 Slac Summer Institute on Particle Physics*, see gr-qc/9811036.

[37] T. C. Mo and C. H. Papas, "Electromagnetic field and wave propagation in Gravitation", *Phys. rev. D* **3**, 1708 (1971).

[38] A. M. Volkov, A.A. Izmest'ev, and G. V. Skrotskii, "The propagation of Electromagnetic Waves in a Riemannian Space", *Soviet Physics JETP* **32**, 636 (1971).

[39] B. Mashhoon, "Scattering of electromagnetic radiation from a black hole", *Phys. Rev. D* **7**, 2807 (1973).

[40] H. P. Robertson and T. W. Noonan, *Relativity and Cosmology*, W. B. Saunders Company, Philadelphia, (1968).

[41] J. Manzano and R. Montemayor, "Propagation of light in a gravitational background", *Phys. Rev. D* **56**, 6378 (1997).

[42] A. M. Anile, "Geometrical optics in general relativity: a study of the higher order corrections", *J. Math. Phys.* **17** , 576 (1976).

[43] J. Bicak and P. Hadrava, "General-relativistic radiative transfer theory in refractive and dispersive media", *Astron. & Astrophys.* **44**, 389-399 (1975).

[44] B. Mashhoon, "Wave propagation in a gravitational field", *Phys. Lett. A* **122**, 299 (1986).

[45] V. Faraoni,"On the rotation of polarization by a gravitational lens", *Astron. Astrophys.* **272**, 385-388 (1993).

[46] B. R. Schupler, R. L. Allshouse, T. A. Clark, "Signal Characteristics of GPS user antennas", *Navigation* **41**, 277 (1994).

[47] C. A. Balanis, *Antenna Theory*, Second Edition, J. Wiley and Sons, Inc., New York (1997).

[48] R. E. Collin and F. J. Zucker, "Antenna Theory, Part I", McGraw-Hill Book Company, New York (1969).

[49] G. Sinclair, "Transmission and Reception of Elliptically Polarized Waves", *Proc. I.R.E.* **38**, 148-151 (1950).

[50] G. H. Price, "On the Relationship Between the Transmitting and Receiving Properties of an Antenna", *IEEE Trans. Ant. Prop.* **AP-34**, 1366-1368 (1986).

[51] See for example, T. B. Bahder, "Pyroelectric Effect in Semiconductor Heterostructures", Superlattices and Microstructures, **14**, 149 (1993).

[52] H. S. Ruse, "Taylor's Theorem in the Tensor Calculus", *Proc. London Math. Soc.* **32**, 87-92 (1931).

[53] H. S. Ruse, "An Absolute Partial Differential Calculus", *Quart. J. Math. Oxford Ser.* **2**, 190 (1931).

[54] J. L. Synge, "A Characteristic Function in Riemannian Space and its Application to the Solution of Geodesic Triangles", *Proc. London Math. Soc.* **32**, 241-258 (1931).

[55] K. Yano and Y. Muto, "Notes on the deviation of geodesics and the fundamental scalar in a Riemannian space", *Proc. Phys.-Math. Soc. Jap.* **18**, 142 (1936).

[56] J. A. Schouten, "Ricci-Calculus. An Introduction to Tensor Analysis and Its Geometrical Applications" (Springer-Verlag, 2nd edition, 1954).

[57] R. W. John, "Zur Berechnung des geodätischen Abstands und assoziierter Invarianten im relativistischen Gravitationsfeld", *Ann. der Phys. Lpz.* **41**, 67-80, (1984).

[58] R. W. John, "The world function of space-time: some explicit exact results for specific metrics", *Ann. der Phys. Lpz.* **41**, 58-70 (1989).

[59] H. A. Buchdahl and N. P. Warner, "On the world function of the Schwarzschild field", *Gen. Rel. and Grav.* **10**, 911-923(1979).

[60] J. M. Gambi, P. Romero, A. San Miguel, and F. Vicente, "Fermi coordinate transformation under baseline change in relativistic celestial mechanics", *International J. Theor. Phys.* **30**, 1097-1116 (1991).

[61] L. I. Sedov, "Inertial navigation equations with consideration of relativistic effects", *Sov. Phys. Dokl.* **21**, 727-729 (1976).

[62] See for example p. 21 in L. D. Landau and E. M. Lifshitz, "Quantum Mechanics: Non Relativistic Theory", Pergamon Press, New York (1977).

[63] F. A. E. Pirani, *Bulletin De L'Academie Polonaise Des Sciences* Cl. III, 1957, Vol. V, No. 2, p. 143–146 (1957).

[64] M. H. Soffel, *Relativity in Astrometry, Celestial Mechanics and Geodesy*, Ch. 3, Springer-Verlag, New York (1989).

[65] V. A. Brumberg, In section 2.3 of *Essential Relativistic Celestial Mechanics*, published under the Adam Hilger imprint by IOP Publishing Ltd, Techno House, Bristol, England (1991).

[66] B. Guinot, "Application of general relativity to metrology", *Metrologia* **34**, 261-290 (1997).

[67] E. Fermi, Atti R. *Accad. Lincei Rend. Cl. Sci. Fis. Mat. Nat.* **31**, 21–23, 51–52, 101–103 (1922).

[68] A. G. Walker, *Proc. Edin. Math. Soc.* **4**, 170 (1935).

[69] F. K. Manasse and C. W. Misner, *J. Math. Phys.* **4**, 735 (1963).

[70] W.-Q. Li and W.-T. Ni *J. Math. Phys.* **20**, 1925 (1975).

[71] N. Ashby and B. Bertotti, *Phys. Rev. D* **34**, 2246 (1986).

[72] W.-Q. Li and W.-T. Ni, *Chin. J. Phys.* **16**, 214 (1978).

[73] W.-T. Ni and M. Zimmermann, *Phys. Rev. D* **17**, 1473 (1978).

[74] W.-Q. Li and W.-T. Ni, *J. Math. Phys.* **20**, 1473 (1979).

[75] T. Fukushima, "The Fermi Coordinate System in the Post-Newtonian Framework", *Celestial Mechanics* **44**, 61 (1988).

[76] R. A. Nelson, J. Math. Phys. **28**, 2379 (1987).; *J. Math. Phys.* **35**, 6224 (1994).

[77] R. A. Nelson, *Gen. Rel. Grav.* **22**, 431 (1990).

[78] T. B. Bahder, "Fermi Coordinates of an Observer Moving in a Circle in Minkowski Space: Apparent Behavior of Clocks", Army Research Laboratory, Adelphi, Maryland, U.S.A., Technical Report ARL-TR-2211, May 2000.

[79] I. P. Vyblyi and N. N. Kostiukovich, "Orbital and Pericentric Reference Systems in the Schwarzschild Field", *Applied Mathematics and Mechanics* **46**, 491-6 (1982).

[80] K.-P. Marzlin,"Fermi Coordinates for Weak Gravitational Fields", *Phys. Rev. D* **50**, 888 (1994).

[81] A. I. Nesterov, "Riemann Normal Coordinates, Fermi Reference System and the Geodesic Deviation Equation", *Classical and Quantum Gravity,* **16**, 465-77 (1999).

[82] See for example, C. W. Misner, K. S. Thorne, and J. A. Wheeler, Gravitation (W. H. Freeman and Company, New York, 1973), pp. 570-583.

[83] For an entertaining description, see D. Sobel, "Longitude", Fourth Estate; ISBN: 1857025717 (1998).

[84] A. S. Eddington, "The Mathematical Theory of Relativity", Chelsea Publishing Company, New York, (1924).

[85] B. W. Parkinson and J. J. Spilker, eds., Global Positioning System: Theory and Applications, vol. I and II (P. Zarchan, editor-in-chief), Progress in Astronautics and Aeronautics, vol. 163 and 164 (Amer. Inst. Aero. Astro., Washington, D.C., 1996).

[86] E. D. Kaplan, *Understanding GPS: Principles and Applications, Mobile Communications Series* (Artech House, Boston, 1996).

[87] B. Hofmann-Wellenhof, H. Lichtenegger, and J. Collins, *Global Positioning System Theory and Practice* (Springer-Verlag, New York, 1993).

[88] See for example, N. Ashby and J. J. Spilker, "Introduction to Relativistic Effects on the Global Positioning System", in *Global Positioning System: Theory and Applications*, B. W. Parkinson and J. J. Spilker, eds., vol. I and II (P. Zarchan, editor-in-chief), *Progress in Astronautics and Aeronautics, vol. 163 and 164* (Amer. Inst. Aero. Astro., Washington, D.C., 1996).

[89] D. Kirchner, "Two-way time transfer via communication satellites", *Proc. IEEE* vol.79, no. 7, 1991.

[90] S. Francis, B. Ramsey, S. Stein, J. Leitner, M. Moreau, R. Burns, R. A. Nelson, T. R. Bartholomew, and A. Gifford, "Time keeping and time dissemination in a distributed space-based clock ensemble", *34th Annual Precise Time and Time Interval (PTTI) Meeting,* December 3-5, Reston, Virginia, USA, 2002.

[91] M. Caputo, *The gravity field of the earth* (Academic Press, N. Y., 1967).

[92] R.R. Bate, D.D. Mueller, and J. E. White, *Fundamental of Astrodynamics*, Pergamon, New York (1971).

INDEX